Master Math: Elementary School Math

By
Kristen Gryskevich

Cengage Learning PTR

Professional • Technical • Reference

Australia, Brazil, Japan, Korea, Mexico, Singapore, Spain, United Kingdom, United States

Professional • Technical • Reference

Master Math:
Elementary School Math
Kristen Gryskevich

Publisher and General Manager, Cengage Learning PTR: Stacy L. Hiquet

Associate Director of Marketing: Sarah Panella

Manager of Editorial Services: Heather Talbot

Senior Marketing Manager: Mark Hughes

Senior Product Manager: Emi Smith

Project and Copy Editor: Kate Shoup

Technical Editor: Megan Hamilton

Interior Layout: Shawn Morningstar

Cover Designer: Mike Tanamachi

Proofreader: Sue Boshers

Printed in the
United States of America
1 2 3 4 5 6 7 16 15 14

Library of Congress Control Number: 2014939186

ISBN-13: 978-1-305-08516-9

ISBN-10: 1-305-08516-7

Cengage Learning PTR
20 Channel Center Street
Boston, MA 02210
USA

Cengage Learning is a leading provider of customized learning solutions with office locations around the globe, including Singapore, the United Kingdom, Australia, Mexico, Brazil, and Japan. Locate your local office at: **international.cengage.com/region.**

Cengage Learning products are represented in Canada by Nelson Education, Ltd.

For your lifelong learning solutions, visit **cengageptr.com.**

Visit our corporate Web site at **cengage.com.**

To my students and parents past and present:

Without all of you,
none of this would have been possible.

Acknowledgments

I want to acknowledge many people who have been by my side through this process.

To my best friends in the world: Thank you for always sending me the gentle reminder to START writing. I needed that nudge on SO many days!

To Megan: My savvy tech editor and teaching partner for 12 years, what would I do without you finishing my thoughts and bouncing my ideas? My teaching is far better because of you.

To Kaci: The math guru of my teaching team, you gave me "little tips" without knowing it and you will always be my math go-to!

To Stacy: Thank you for being my biggest cheerleader and making me feel smarter than I thought I was. You had faith in me from the beginning.

To Lori: You are such an extraordinary mentor and teacher. You make me want to get better every day. Your passion is contagious!

To everyone at Cengage Learning: Thanks for deciphering what I wanted to say and making it look readable. You make me look so professional!

To the students of my 2013–2014 class: Thank you for showing excitement about your teacher being an author! You helped me all year to make the steps of the algorithms clear for my reader. You are AWESOME!

To Mom and Dad: You both always told me that I could do anything I set my hands on and I never doubted that. Your love has guided me and has given me the courage to do things I didn't imagine possible! I love you to the moon and back!

To Ali: You are so much more than a sister. You are my ROCK! You have always made me believe that teaching is the most important job in the world. Thank you for believing in me and being my #1 fan. Words can't express how much I love you.

To my boys: All three of you never questioned why dinner wasn't on the table and laundry wasn't done. You gave me this time and I am forever grateful. You keep me on my toes. You are my loves!

What would I do without ALL of you in my life? I love each and every one of you.

Kris xx

About the Author

Kristen Gryskevich developed her love for teaching at West Virginia University, where she received her bachelor of science degree. She has also earned a master's degree from Indiana Wesleyan University.

Kristen has been teaching elementary age students for 19 years. She spent the first three years of her teaching career in the Indianapolis Public School district, teaching third grade. She has taught many different upper-grade levels, including third, fourth, and sixth, and is currently a fifth-grade teacher in the Brownsburg School Corporation in Brownsburg, Indiana. Kristen can attribute much of her students' success to the training and support she has received from administrators and colleagues in Brownsburg. Seeing parents struggle with helping their children with new strategies was most definitely her inspiration for this book.

Kristen is a board member of the Brownsburg Education Foundation, helping to raise money and give back to both students and teachers with grants and scholarships. In her spare time, Kristen enjoys running, reading, and spending time with her husband, Mike, and two sons, Parker and Caden. Kristen has run half marathons and loves to run theme 5Ks with her friends. In the summer, you can find her at a baseball field or venturing around Indianapolis on adventures with her boys. But most days, you will find her in the classroom, helping to make learning fun for her always entertaining and busy fifth graders!

To the Reader

The fact that math has changed over the years is no secret to parents. Are you frustrated that you can't help your child because the way that your child is learning math is not the way that you learned it? Whether they are learning "Everyday Math" or another math series, the way students learn math is changing.

This book is the key to making math easier in your home! This is a concise guide on the different algorithms, with strategies to help you make your child a math success.

In the first few chapters, you will learn different mathematical algorithms (that is, the steps you must follow to get to the solution) that your child may be using to solve addition, subtraction, multiplication, and division problems. There are many different kinds of algorithms; every child will be able to find one that will help him or her be successful with computation. For each algorithm, there are step-by-step examples and practice problems to help you tackle them like a pro. If your child is struggling with computation, I promise you will be able to find one way that makes sense.

The other chapters give you tips on how to help your child with fraction, decimal, percent, measurement, and algebra concepts. I will give you suggestions on how to make math concepts easier for your child, sharing tricks I teach in my classroom to help my students excel. There is no one best way to do math; in this book, I hope you will find new ways that may help concepts click with your child.

I hope that you will use this book as a reference when helping your child with homework. I am confident that with it, you will experience less frustration and tears and more smiles. Math is easy when you're given tools that make sense! Get ready, because you have just found the key to success for your child.

Table of Contents

Chapter 1

Addition Algorithms

Traditional Addition

Traditional addition is the way we were all taught, right? You are probably thinking that's the only way you know how to add numbers! When adding traditionally, the number is added from right to left, one place at a time, carrying numbers if required. If traditional addition is a struggle for your child, however, this approach is not your only option. Many children have a hard time remembering to carry their numbers, or they carry the wrong number to the next place value. Fortunately there are many ways for children to master this computation process. If your child is struggling, try another algorithm and see if it makes it easier!

The Steps to This Algorithm

1. Line up the numbers you are adding by their place value.
2. Begin at the right of the problem, or the smallest place value.
3. Add the ones and regroup (carry) if necessary. Continue to add, moving left.
4. Write the answer for each place under the correct column of the problem.

Make sure your child is adding one column at a time and writing the answer directly under it. If your child is required to carry, make sure that when looking at the problem, he or she can see the answer. This will ensure that the correct number is being carried. This is a common mistake for many children.

$$\begin{array}{r} {}^{1} \\ 43 \\ +29 \\ \hline 2 \end{array}$$

says "12."

Examples

289 + 474 = 763

$$\begin{array}{r} {}^{1} \\ 289 \\ +474 \\ \hline 3 \end{array}$$

- add the ones (9 + 4 = 13 ones)
- regroup (13 ones = 1 ten + 3 ones)

$$\begin{array}{r} {}^{1}\,{}^{1} \\ 289 \\ +474 \\ \hline 63 \end{array}$$

- add the tens (8 + 7 + 1= 16 tens)
- regroup (16 tens = 1 hundred + 6 tens)

$$\begin{array}{r} {}^{1}\,{}^{1} \\ 289 \\ +474 \\ \hline 763 \end{array}$$

- add the hundreds (2 + 4 + 1 = 7 hundreds)
- 763 is the final answer

4,478 + 854 = 5,332

```
    1
 4,478
+  854
------
     2
```

- line up the numbers by place value
- add the ones (8 + 4 = 12 ones)
- regroup (12 ones = 1 ten + 2 ones)

```
   1 1
 4,478
+  854
------
    32
```

- add the tens (7 + 5 + 1 = 13 tens)
- regroup (13 tens = 1 hundred + 3 tens)

```
  1 1 1
 4,478
+  854
------
   332
```

- add the hundreds (4 + 8 + 1 = 13 hundreds)
- regroup (13 hundreds = 1 thousand + 3 hundreds)

```
  1 1 1
 4,478
+  854
------
 5,332
```

- add the thousands (4 + 1 = 5 thousand)
- 5,332 is the final answer

Practice Problems

1.1 253 + 415 =

1.2 408 + 314 =

1.3 $7{,}452 + 1{,}475 =$

1.4 $2{,}478 + 852 =$

1.5 $14{,}758 + 2{,}854 =$

Partial Sums

If your child is struggling with traditional addition, this algorithm may work as a visual aid to help show your child the process of adding numbers. Partial sums is another method to help solve multi-digit problems. This strategy is different because it allows children to break the problem into separate parts and add each part mentally. The only drawback to this strategy is that children must have a solid foundation of place value and knowledge of the value of the digit from where it is placed in a number. In simple terms, using this approach, a child will add one place value column at a time, and then add the partial sums together to get the solution. This helps many children who struggle with remembering to carry or regroup their digits to the next place. This strategy has helped many of my students become successful adding numbers even into the millions.

The Steps to This Algorithm

1. Write the problem, lining up the digits by place value.
2. Add the digits in each place value, starting with the digits with the greatest value.
3. Add the partial sums that you got together to get your solution.

> When starting, the most important step is making sure that the problem is lined up by place value. That is, the hundreds will be in a straight line, then tens in a straight line, and so on. Draw lines between the places to help children visualize the place to add together.

Examples

246 + 132 = 378

2	4	6
+1	3	2

First, add the numbers in the hundreds place. It is very important that children understand that they are adding 200+100 and not 2+1 because the numbers are in the hundreds place of the number.

2	4	6	
+**1**	3	2	
3	**0**	**0**	(200 + 100 = 300)

Next, comes the tens place. Again, children need to understand that they are adding digits in the tens place.

2	**4**	6	
+1	**3**	2	
3	0	0	(200 + 100 = 300)
	7	**0**	(40 + 30 = 70)

Finally, add the ones column together.

2	4	**6**	
+1	3	**2**	
3	0	0	(200 + 100 = 300)
	7	0	(40 + 30 = 70)
		8	(6 + 2 = 8)

Once you have done all the partial sums, you add the numbers together to get the solution.

2	4	6	
+1	3	2	
3	0	0	
	7	0	• add (300 + 70 + 8 = 378)
		8	• 378 is the final answer
3	7	8	

Work on this strategy with problems that do not require the places to carry to the next. This will ensure that children master the steps to the algorithm before they move to harder problems. Practice problems 1.6–1.10 will help you teach the strategy to your child and help you ensure they understand the strategy before moving on.

Here is one more problem before moving on to problems that require children to move the digits into the next column because the value is greater than one digit.

635 + 243 =

6	3	5
+2	4	3

First, add the numbers in the hundreds place. Again stress the fact that you are adding 600 + 200 and not 6 + 2.

6	3	5	
+**2**	4	3	
8	**0**	**0**	(600 + 200 = 800)

Next comes the tens place. You will be adding 30 + 40.

6	**3**	5	
+2	**4**	3	
8	0	0	(600 + 200 = 800)
	7	**0**	(30 + 40 = 70)

Finally, add the ones column, 5 + 3.

6	3	**5**	
+2	4	**3**	
8	0	0	(600 + 200 = 800)
	7	0	(30 + 40 = 70)
		8	(5 + 3 = 8)

Once you have done all the partial sums, add the numbers together to get the solution.

6	3	5	
+2	4	3	
8	0	0	
	7	0	• add (800 + 70 + 8 = 878)
		8	• 878 is the final answer
8	7	8	

You might be wondering what happens when the value of the number is larger than one digit. The key to this is making sure that the problem is lined up and that lines are drawn between places to keep the problem organized. Here's an example:

745 + 264 = 1,009

7	4	5
+2	6	4

You can see by looking at the problem that (60 + 40) in the tens place is going to add up to 100, which will need to be written in the hundreds place. This might confuse your child at first. First, add the numbers in the hundreds place (700 + 200).

7	4	5	
+**2**	6	4	
9	**0**	**0**	(700 + 200 = 900)

Next comes the tens place. Again, children need to understand that they are adding digits in the tens place. However, this time the answer is greater than tens and needs to be written in the hundreds place. Notice the arrow in the problem. This shows you how to write the partial sum of (40 + 60) in the problem.

7	**4**	5	
+2	**6**	4	
9	0	0	(700 + 200 = 900)
1	**0**	**0**	(40 + 60 = 100)

Finally, add the ones column together.

7	4	**5**	
+2	6	**4**	
9	0	0	(700 + 200 = 900)
1	0	0	(40 + 60 = 100)
		9	(5 + 4 = 9)

Once you have done all the partial sums, you add the numbers together to get the solution (900 + 100 + 9 = 1,009).

7	4	5	
+2	6	4	
9	0	0	
1	0	0	• add (900 + 100 + 9 = 1,009)
		9	• 1,009 is the final answer
1,0	0	9	

You should be familiar with the steps at this point. You follow the same steps for any problem that you need to solve. Here are a few more examples that will help you visualize how to solve the problem if you get stuck.

842 + 239 = 1,081

	8	4	2	
+	2	3	9	
1	0	0	0	(800 + 200 = 1,000)
		7	0	(40 + 30 = 70)
		1	1	(2 + 9 − 11)
1,	0	8	1	(1,000 + 70 + 11 = 1,081)

931 + 850 = 1,781

	9	3	1	
+	8	5	0	
1	7	0	0	(900 + 800 = 1,700)
		8	0	(30 + 50 = 80)
			1	(1 + 0 = 1)
1,	7	8	1	(1,700 + 80 + 1 = 1,781)

1,485 + 574 = 2,059

1	4	8	5	
+	5	7	4	
1	0	0	0	(1,000 + 0 = 1,000)
	9	0	0	(400 + 500 = 900)
	1	5	0	(80 + 70 = 150)
			9	(5 + 4 = 9)
2,	0	5	9	(1,000 + 900 + 150 + 9 = 2,059)

24,806 + 42,852 = 67,658

2	4,	8	0	6	
+4	2,	8	5	2	
6	0	0	0	0	(20,000 + 40,000 = 60,000)
	6	0	0	0	(4,000 + 2,000 = 6,000)
	1	6	0	0	(800 + 800 = 1,600)
			5	0	(0 + 50 = 50)
				8	(6 + 2 = 8)
6	7,	6	5	8	

↖(60,000 + 6,000 + 1,600 + 50 + 8 = 67,658)

Practice Problems

1.6 $253 + 415 =$

1.7 $145 + 553 =$

1.8 $514 + 285 =$

1.9 $336 + 252 =$

1.10 $843 + 56 =$

1.11 $548 + 789 =$

1.12 1,568 + 946 =

1.13 5,758 + 8,327 =

1.14 12,741 + 25,369 =

1.15 85,247 + 45,021 =

Column Addition

Column addition is another algorithm that focuses on adding the numbers in each place value and then adjusting the columns as necessary. The only difference is that when adding, the sum of each column is placed under that place value and then adjusted instead of having to carry or regroup. This will help children visualize what you are doing when you regroup.

The Steps to This Algorithm

1. Line up the numbers by their place value and draw lines between the places.
2. Add the numbers in each column one at a time.
3. Adjust the place value as needed.

Examples

456 + 282 = 738

4	5	6	
+2	8	2	• add the numbers in each column
6	13	8	

4	5	6	
+2	8	2	• adjust the place value as needed
6	**13**	8	• adjust the tens
7 ←	**3**	8	• 738 is the final answer
7	3	8	

878 + 598 = 1,476

8	7	8	
+5	9	8	
13	16	16	• add the columns
14 ←	6	16	• adjust the tens
1,4	7 ←	6	• adjust the ones
1,4	7	6	• 1,476 is the final answer

1,436 + 2,896 = 4,332

1,	4	3	6	
+2,	8	9	6	
3	12	12	12	• add the columns
4 ←	2	12	12	• adjust the hundreds
4	3 ←	2	12	• adjust the tens
4	3	3 ←	2	• adjust the ones
4,	3	3	2	• 4,332 is the final answer

Practice Problems

1.16 $506 + 754 =$

1.17 $347 + 540 =$

1.18 $1{,}267 + 839 =$

1.19 $34{,}643 + 8{,}902 =$

1.20 $67{,}309 + 5{,}391 =$

If your child is struggling with addition and, more specifically, place value, review the place with students prior to moving on with addition.

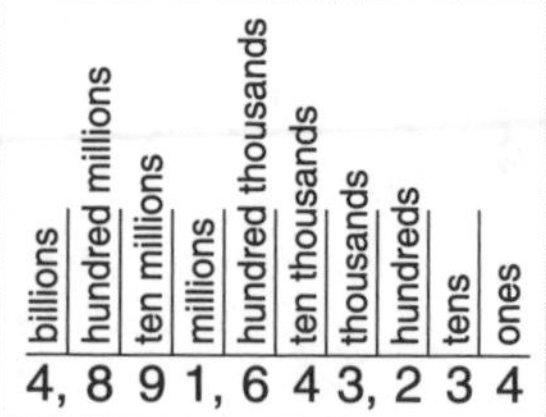

billions	hundred millions	ten millions	millions	hundred thousands	ten thousands	thousands	hundreds	tens	ones
4,	8	9	1,	6	4	3,	2	3	4

Traditional Addition with Decimals

This algorithm is done the same way as traditional addition. The digits are lined up with their place value by aligning the decimal places in both numbers. The numbers are then added from right to left, one digit at a time, regrouping when necessary.

The Steps to This Algorithm

1. Line up the two numbers by their place value, paying close attention to ensure the decimal points are aligned.
2. Carry the decimal point straight down before starting any computation.
3. Add from right to left, one digit at a time, and regroup when necessary.

> Many children will line up the numbers to the left or the right, not paying attention to the decimal. Use graph paper and place the decimal in the problem before writing the numbers. Then, have your child fill in the whole numbers to the left of the number and the decimals to the right of the number. I also have the children in my classroom highlight the decimals before starting a problem to make sure they don't forget to place them in the problem.

Examples

$4.67 + 6.34 = 11.01$

First, line up the two numbers by their place value and decimal point and carry down the decimal point immediately.

$$\begin{array}{r} 4.67 \\ +6.34 \\ \hline . \end{array}$$

Then, add the numbers from right to left, regrouping when necessary.

```
   1
  4.67
 +6.34     • add the hundredths
 -----       (0.07 + 0.04 = 0.11)
   . 1     • regroup
```

```
  1 1
  4.67
 +6.34     • add the tenths
 -----       (0.6 + 0.3 + 0.1 = 1.0)
   .01     • regroup
```

```
  1 1
  4.67
 +6.34     • add the ones
 -----       (4 + 6 + 1 = 11)
 11.01 <-• 11.01 is the final answer
```

0.456 + 2.86 = 3.316

```
  0.456
 +2.860
 ------
   .
```

Notice on this problem that there is not a number in the thousandths place of the bottom number. I always tell the children in my classroom to place a zero in that place so that there is never confusion.

Follow all the preceding steps, but add a zero to make the bottom number 2.860. This does not change the value of the number.

```
  1 1
  0.456
 +2.860
 ------
  3.316
```

4.56 + 2.378 = 6.938

First, carry the decimal down and then place a zero in the thousandths to make the top number 4.560.

```
  4.560
 +2.378
 ------
   .
```

Follow the same steps as before to get the sum.

$$\begin{array}{r} \overset{1}{} \\ 4.560 \\ +2.378 \\ \hline 6.938 \end{array}$$

Practice Problems

1.21 3.4 + 7.6 =

1.22 0.8 + 3.2 =

1.23 78.5 + 5.4 =

1.24 46.65 + 34.96 =

1.25 0.007 + 0.569 =

Partial Sums with Decimals

This algorithm uses the same approach as partial sums, adding each place value column one at a time moving from left to right, or the largest value numbers to the smallest. Children need to make sure to line up the decimal point in both numbers.

The Steps to This Algorithm

1. Line up the numbers by their decimal point and place value.
2. Add each place value one at a time, starting with the largest value number (left to right).
3. Add all the partial sums to get the final sum.

Examples

5.67 + 3.84 = 9.51

5.	6	7	
+**3.**	8	4	
8.	**0**	**0**	(5 + 3 = 8.00)

5.	**6**	7	
+3.	**8**	4	
8.	0	0	
1.	**4**	**0**	(0.6 + 0.8 = 1.40)

5.	6	**7**	
+3.	8	**4**	
8.	0	0	
1.	4	0	
0.	**1**	**1**	(0.07 + 0.04 = 0.11)

5.	6	7
+3.	8	4
8.	0	0
1.	4	0
0.	1	1
9.	5	1

• add (8.00 + 1.40 + 0.11 = 9.51)

• 9.51 is the final answer

0.87 + 2.764 = 3.634

0.	8	7	0	← • add a zero.
+2.	7	6	4	
2.	0	0	0	(0 + 2 = 2.000)
1.	5	0	0	(0.8 + 0.7 = 1.500)
0.	1	3	0	(0.07 + 0.06 = 0.130)
0.	0	0	4	(0.000 + 0.004 = 0.004)
3.	6	3	4	

• add 2 + 1.5 + 0.13 + 0.004

• 3.634 is the final answer

Practice Problems

1.26 5.76 + 4.2 =

1.27 0.68 + 0.54 =

1.28 $7 + 4.567 =$

1.29 $3.42 + 6.9 =$

1.30 $9.68 + 18.44 =$

Chapter 2

Subtraction Algorithms

Traditional Subtraction

This is the algorithm that you will be familiar with from learning math yourself in elementary school. You subtract from right to left and regroup (borrow) as needed throughout the problem.

Many children struggle to remember to cross off the original number in the next place value when regrouping. This is a very common mistake with children in my classroom. I always tell my students that if you are going to "borrow" something from your neighbor, you need to make sure you take what you need. This helps them to remember to cross off the original number and make it one less when regrouping.

The Steps to This Algorithm

1. Align your numbers by place from right to left, lining up digits with the same place value.
2. Start with the ones column, subtract one column at a time, and regroup as needed.

I always have my students tell me this rhyme to help decide whether you can subtract or if you need to regroup (borrow): "More on top, stop. Subtract. More on the floor, go next door, and borrow one more."

Examples

654 – 329 = 325

The first step is to decide whether you can subtract 9 ones from 4 ones. You cannot, so you need to go to the tens place and regroup, or borrow 1 ten (14 ones – 9 ones = 5 ones).

```
  4 14
6 5̸ 4̸     subtract the ones place
– 3 2 9 ←"more on the floor,
      5     so go next door"
```

Next, can you subtract 2 tens from 4 tens? Yes, you can. There's more on top, so stop and subtract (4 tens – 2 tens = 2 tens).

```
  4 14
6 5̸ 4̸     subtract the tens place
– 3 2 9 ←"more on top, so STOP!"
    2 5
```

Finally, can you subtract 3 hundreds from 6 hundreds? Yes, you can. So, stop and subtract (6 hundred – 3 hundred = 3 hundred).

```
  4 14
6 5̸ 4̸     subtract the hundreds place
– 3 2 9 ←"more on top, so STOP!"
  3 2 5
```

6,294 – 885 = 5,409

The first step is to line up the digits by their place value from right to left. You need to be sure to line up correctly because you are subtracting a three-digit number from a four-digit number.

thousands	hundreds	tens	ones
6	2	9	4
–	8	8	5

The second step is to think, can you subtract 5 ones from 4 ones? No, you cannot. So you need to go to the tens place and regroup, or borrow 1 ten (14 ones – 5 ones = 9 ones).

```
      8 14
  6,2 9 4
–   8 8 5
---------
        9
```

Next, can you subtract 8 tens from 8 tens? Yes, you can. So, stop and subtract (8 tens – 8 tens = 0 tens).

```
      8 14
  6,2 9 4
–   8 8 5
---------
      0 9
```

Move to the hundreds place in the number. Can you subtract 8 hundreds from 2 hundreds? No, so you need to go to the thousands and regroup, or borrow 1 thousand (12 hundreds – 8 hundreds = 4 hundreds).

```
  5 12 8 14
  6, 2 9 4
–    8 8 5
----------
     4 0 9
```

Finally, there is nothing to subtract from the 5 thousands, so carry that number down to your answer. You are finished.

$$\begin{array}{r} \overset{5}{\cancel{6}},\overset{12}{\cancel{2}}\,\overset{8}{\cancel{9}}\,\overset{14}{\cancel{4}} \\ -\quad 8\,8\,5 \\ \hline \mathbf{5},4\,0\,9 \end{array}$$

Your child might be struggling with problems that include a 0 because he or she doesn't understand how to go over to the next place to regroup. This issue might be revealed by your child noting, "I can't regroup from 0!" Reassure your child that you can. The steps follow.

500 – 323 = 177

The first step is to decide whether you can subtract 3 ones from 0 ones. No, you cannot. So you need to go to the tens place and regroup, or borrow 1 ten. However, in this problem, you cannot borrow from 0, so you need to move another place, to the hundreds, and borrow 1 hundred.

$$\begin{array}{r} \overset{4}{\cancel{5}}\,\overset{10}{\cancel{0}}\,0 \\ -\,3\,2\,3 \\ \hline \end{array}$$

Next, you need to borrow from the tens for the ones place. So, regroup, or borrow 1 ten to give to the ones place and then subtract (10 ones – 3 ones = 7 ones).

$$\begin{array}{r} \overset{4}{\cancel{5}}\,\overset{\overset{9}{\cancel{10}}}{\cancel{0}}\,\overset{10}{\cancel{0}} \\ -\,3\,2\,3 \\ \hline 7 \end{array}$$

Then, subtract the tens place (9 tens – 2 tens = 7 tens).

$$\begin{array}{r} \overset{4}{\cancel{5}}\,\overset{\overset{9}{\cancel{10}}}{\cancel{0}}\,\overset{10}{\cancel{0}} \\ -\,3\,\mathbf{2}\,3 \\ \hline \mathbf{7}\,7 \end{array}$$

Finally, subtract the hundreds (4 hundred – 3 hundred = 1 hundred).

```
    9
 4 10 10
 5  0  0
– 3  2  3
  1  7  7
```

If your child is struggling with remembering to regroup when there are zeroes in the problem, here is a shortcut you can use. Many students in my room use this strategy to help them solve problems across zeroes.

404 – 296 = 108

The first step is to draw a circle around the zero or zeros and the number preceding it. In this case, you would circle 40 in 404. Then borrow 1 from the 40 and make it 39. Cross out the 40 and write 39 above it. Then, make the 4 into a 14. Your child will quickly catch on that this is what the problem looks like when you do traditional regrouping, but with a shortcut!

```
  3 9 14
  4 0 4
– 2 9 6
```

Finally, subtract all the places starting with the ones place.

```
  3 9 14
  4 0 4
– 2 9 6
  1 0 8
```

Here is another example, with numbers in the thousands.

9,006 – 4,797 = 4,209

The first step is to look at the problem and determine whether you will need to regroup. In this case, you need to begin by regrouping for the ones place. Circle the 900 in 9,006. Borrow 1 from the 900 and make it 899. Cross out the 900 and write 899 above it. Then, make the 6 into a 16.

$$\begin{array}{r} \scriptstyle 8\ \ 9\ \ 9\ \ 16 \\ 9,0\,0\,\not{6} \\ -4,7\,9\,7 \\ \hline \end{array}$$

Finally, subtract all the places, starting with the ones place.

$$\begin{array}{r} \scriptstyle 8\ \ 9\ \ 9\ \ 16 \\ 9,0\,0\,\not{6} \\ -4,7\,9\,7 \\ \hline \mathbf{4,2\,0\,9} \end{array}$$

Practice Problems

2.1 542 – 39 =

2.2 875 – 458 =

2.3 478 – 288 =

2.4 5,245 – 857 =

2.5 $6{,}452 - 4{,}388 =$

2.6 $508 - 459 =$

2.7 $800 - 287 =$

2.8 $1{,}000 - 562 =$

2.9 $5{,}004 - 2{,}458 =$

2.10 $7{,}000 - 4{,}127 =$

Trade First Subtraction

This algorithm looks very much like tradition subtraction with one main difference: You do all the regrouping before any subtracting of numbers begins. If you are using the shortcut for subtracting with zeroes, then this will be an easy method for your child to grasp. This method allows for fewer errors in regrouping because it is done all at once.

The Steps to This Algorithm

1. Look at the top number in each place value column. If the number on top is smaller than the bottom number, then it is necessary to regroup/borrow.
2. Regroup/borrow numbers as necessary, starting from the ones and moving to higher places. Do all necessary regrouping first; do not do any of the computation/subtracting of the numbers.
3. When you have done all the regrouping, subtract the numbers starting in the ones and moving to larger value places.

Examples

782 – 125 = 657

The first step is to line up the digits by their place value from right to left.

$$\begin{array}{r} 782 \\ -125 \\ \hline \end{array}$$

The second step is to think, can you subtract 5 ones from 2 ones? No, so you need to go to the tens and place and regroup, or borrow 1 ten (8 tens – 1 ten = 7 tens). Give the 10 that you borrowed to the ones and make it a 12.

$$\begin{array}{r} {\scriptstyle 7\;12} \\ 7\not{8}\not{2} \\ -125 \\ \hline \end{array}$$

Next, think can you subtract 2 tens from 7 tens? Yes, you can. So stop. There is no need to do any other regrouping. In this case, it is now time to subtract because you have finished all the regrouping necessary for this problem.

$$\begin{array}{rrrr} & & {\scriptstyle 7} & {\scriptstyle 12} \\ & 7 & \not{8} & \not{2} \\ - & 1 & 2 & 5 \\ \hline & 6 & 5 & 7 \end{array}$$

2,456 – 787 = 1,669

The first step is to line up the digits by their place value from right to left. You need to be sure to line them up correctly because you are subtracting a three-digit number from a four-digit number.

$$\begin{array}{rrrrr} & 2, & 4 & 5 & 6 \\ - & & 7 & 8 & 7 \\ \hline \end{array}$$

The second step is to think, can you subtract 7 ones from 6 ones? No, so you need to go to the tens place and regroup, or borrow 1 ten for 10 ones (5 tens – 1 ten = 4 tens). Give the 10 ones that you borrowed to the ones place and make it a 16.

$$\begin{array}{rrrrr} & & & {\scriptstyle 4} & {\scriptstyle 16} \\ & 2, & 4 & \not{5} & \not{6} \\ - & & 7 & 8 & 7 \\ \hline \end{array}$$

Next, look at the tens place. Can you subtract 8 tens from 4 tens? No, so you need to go to the hundreds place and regroup, or borrow 1 hundred for 10 tens (4 hundreds – 1 hundred = 3 hundreds). Give the 10 tens you borrowed to the tens place and make it a 14.

$$\begin{array}{rrrrr} & & {\scriptstyle 3} & {\scriptstyle 14} & {\scriptstyle 16} \\ & 2, & \not{4} & \not{5} & \not{6} \\ - & & 7 & 8 & 7 \\ \hline \end{array}$$

Then, look at the hundreds place. Can you subtract 7 hundreds from 3 hundreds? No, so you need to go over to the thousands place and regroup, or borrow 1 thousand for 10 hundreds (2 thousands – 1 thousand = 1 thousand). Give the 10 hundreds to the hundreds place and make the 3 a 13.

$$\begin{array}{r} {}^{1}\cancel{2},{}^{13}\cancel{4}\,{}^{14}\cancel{5}\,{}^{16}\cancel{6} \\ -\quad 7\,8\,7 \\ \hline \end{array}$$

You have finished all the regrouping necessary, so the last step is to subtract starting in the ones place and moving right to left.

$$\begin{array}{r} {}^{1}\cancel{2},{}^{13}\cancel{4}\,{}^{14}\cancel{5}\,{}^{16}\cancel{6} \\ -\quad 7\,8\,7 \\ \hline \mathbf{1,6\,6\,9} \end{array}$$

Practice Problems

2.11 542 – 478 =

2.12 815 – 726 =

2.13 1,452 – 874 =

2.14 4,421 – 387 =

2.15 7,213 – 1,607 =

Partial-Differences Subtraction

This strategy is different from the others because the computation is done from left to right. You start at the place with the greatest value and subtract the lesser number from the larger number. You follow these steps until the final difference is calculated. This algorithm would be good for children who have a solid grasp on math facts and place value concepts.

The Steps to This Algorithm

1. Subtract from left to right, one place value (column) at a time, until you get to the smallest place. Subtract the subtrahend (bottom number) from the minuend (top number). Sometimes, the larger number is on the bottom; in that case, you will get a negative number.
2. Find the total by adding or in some cases subtracting the answer from each column.

Examples

852 – 441 = 411

The first step is to rename 852 as 800 + 50 + 2 and 441 as 400 + 40 + 1.

```
 852   (800 + 50 + 2)
−441   (400 + 40 + 1)
────
```

Then, start with the place that has the largest value and subtract from left to right. In this problem subtract starting with the numbers in the hundreds (800 – 400 = 400; 50 – 40 = 10; and 2 – 1 = 1).

The final step is to find the total ([400 + 10 + 1] = 411).

$$\begin{array}{r} 852 \\ -441 \\ \hline \end{array}$$

hundreds (800 – 400 = 400)
tens (50 – 40 = 10)
ones (2 – 1 = 1)
(400 + 10 + 1 = 411)

8,417 – 4,274 = 4,143

The first step is to rename 8,417 as 8,000 + 400 + 10 + 7 and 4,274 as 4,000 + 200 + 70 + 4.

$$\begin{array}{r} 8,417 \\ -4,274 \\ \hline \end{array}$$

(8,000 + 400 + 10 + 7)
(4,000 + 200 + 70 + 4)

Then, start with the place that has the largest value and subtract from left to right. In this problem subtract starting with the numbers in the thousands.

The final step is to find the total ([4,000 + 200 – 60 + 3] = 4,143).

$$\begin{array}{r} 8,417 \\ -4,274 \\ \hline \end{array}$$

thousands (8,000 – 4,000 = 4,000)
hundreds (400 – 200 = 200)
tens (10 – 70 = – 60)
ones (7 – 4 = 3)
(4,000 + 200 – 60 + 3 = 4,143)

16,423 – 4,852 = 11,571

Remember, the first step is to rename 16,423 as 10,000 + 6,000 + 400 + 20 + 3 and 4,852 as 4,000 + 800 + 50 + 2.

$$\begin{array}{r l} 16{,}423 & (10{,}000 + 6{,}000 + 400 + 20 + 3) \\ -\ \ 4{,}852 & (4{,}000 + 800 + 50 + 2) \\ \hline \end{array}$$

Then, start with the place that has the largest value and subtract from left to right. In this problem subtract starting with the numbers in the ten thousands.

The final step is to find the total ([10,000 + 2,000 – 400 – 30 + 1] = 11,571).

$$\begin{array}{r l} & \text{ten thousands } (10{,}000 - 0) = 10{,}000 \\ & \text{thousands } (6{,}000 - 4{,}000) = 2{,}000 \\ & \text{hundreds } (400 - 800) = -400 \\ 16{,}423 & \text{tens } (20 - 50) = -30 \\ -\ \ 4{,}852 & \text{ones } (3 - 2) = 1 \\ \hline 11{,}571 & 10{,}000 + 2{,}000 - 400 - 30 + 1 = 11{,}571 \end{array}$$

Practice Problems

2.16 648 – 437 =

2.17 705 – 654 =

2.18 1,341 – 630 =

2.19 5,746 – 1,697 =

2.20 21,543 – 17,462 =

Traditional Subtraction with Decimals

This algorithm is identical to the steps with subtracting without decimals. If your child is not having a problem with subtracting in this way, then this is the way you will want to reinforce computation with decimals. The computation steps are the same after you line up the decimals and numbers according to place value.

The Steps to This Algorithm

1. Align your numbers by place from right to left, lining up digits with the same place value.
2. Carry the decimal straight down before doing any computation.
3. Start with the ones column, subtract one column at a time, and regroup as needed.

Examples

5.4 – 3.78 = 1.62

First, line up the two numbers by their place value and decimal point, and carry down the decimal point immediately. Notice that there is not a zero in the hundredths place in the first number (5.4). You must add a zero before subtracting.

$$\begin{array}{r} 5.4\mathbf{0} \\ -3.78 \\ \hline . \end{array}$$

• change 5.4 to 5.40
• carry the decimal down

Then, subtract the numbers from right to left, regrouping when necessary.

```
    3 10
 5.4 0     • subtract the hundredths
-3.7 8     • you need to regroup
 ------
  .  2
```

```
 4 13 10
 5.4 0     • subtract the tenths
-3.7 8     • you need to regroup
 ------
  .6 2
```

```
 4 13 10
 5.4 0     • subtract the ones
-3.7 8     • 1.62 is the final answer
 ------
 1.6 2
```

If there is an open place value in any part of the number, have your child place a zero automatically. This will help your child remember that you always need to have something to subtract from in a problem. This step is especially important when there is not a number in one of the places in the top number.

54.7 – 7.63 = 47.07

First, line up the numbers by their place value, aligning the decimal points, and carry the decimal down to the answer. Then, change 54.7 to 54.70 to make both numbers have the same number of places.

```
  54.70    • change 54.7 to 54.70
-  7.63    • carry the decimal down
 ------
    .
```

Then, subtract from right to left one place at a time and regroup as necessary.

$$\begin{array}{r} {}^{4}\cancel{5}\,{}^{14}\cancel{4}.{}^{6}\cancel{7}\,{}^{10}\cancel{0} \\ -\quad 7.63 \\ \hline 47.07 \end{array}$$

- subtract from right to left one place at a time
- regroup as necessary
- 47.07 is the final answer

$34.6 - 27.345 = 7.255$

$$\begin{array}{r} {}^{2}\cancel{3}\,{}^{14}\cancel{4}.{}^{5}\cancel{6}\,{}^{9}\cancel{10}\!\!\cancel{0}\,{}^{10}\cancel{0} \\ -27.345 \\ \hline 7.255 \end{array}$$

Practice Problems

2.21 $9.2 - 4.8 =$

2.22 $14.7 - 6.4 =$

2.23 $45.6 - 23.31 =$

2.24 $4.37 - 2.053 =$

2.25 $28 - 21.88 =$

Trade First with Subtraction of Decimals

This algorithm is perfect for a child who is struggling to regroup correctly. This method does all the regrouping before doing any of the computation. When subtracting with decimals, it is important to carry the decimal down immediately before doing any work on the problem to ensure that this step is completed.

The Steps to This Algorithm

1. Line up the numbers by the decimal place and then add zeroes to make both numbers have the same number of places.
2. Look at the top number in each place value column. If the number on top is smaller than the bottom number, then it is necessary to regroup/borrow.
3. Regroup/borrow numbers as necessary, starting from the ones and moving to higher places. Do all necessary regrouping first; do not do any of the computation/subtracting of the numbers.
4. When you have done all the regrouping, subtract the numbers starting in the ones and moving to larger value places.

Examples

8 – 4.56 = 3.44

The first step is to line up the number by the decimal place. Then, carry the decimal down immediately. Add zeroes as necessary to make the numbers have the same number of decimal places. In this case, you need to change 8 to 8.00.

```
 8.00
-4.56
-----
  .
```

Next, do all the regrouping before doing any computation. Because you are subtracting across zeroes, use the shortcut from the section "Traditional Subtraction" at the beginning of the chapter. The first step is to look at the problem. In this case, you will circle the 80 in 8.00.

Borrow 1 from the 80 and make it 79. Cross out the 80 and write 79 above it. Then, make the 0 into a 10.

$$\begin{array}{r} \overset{7\ \ 9\ \ 10}{\boxed{8.0}\,0} \\ -4.56 \\ \hline . \end{array}$$

Finally, subtract the numbers in each column to get the final answer.

$$\begin{array}{r} \overset{7\ \ 9\ \ 10}{\boxed{8.0}\,0} \\ -4.56 \\ \hline 3.44 \end{array}$$

Practice Problems

2.26 14.7 – 0.9 =

2.27 18.3 – 8.72 =

2.28 74.84 – 14.364 =

2.29 345.8 – 73.36 =

2.30 2.2 – 1.836 =

Chapter 3

Multiplication Algorithms

Traditional Multiplication

Traditional multiplication is probably the way you learned to do multiplication when you were young. Presumably, this is the one way that you know how to multiply and you use it when needed. Solving a multiplication problem in the traditional manner, you move from right to left, multiplying as you go and regrouping as necessary. Often, however, children forget to multiply one of the numbers in a larger problem or to line up the numbers under the correct place value. One thing I see all the time in my own classroom is that children do not write the larger number on top as they multiply. This results in the wrong answer. Writing the problem correctly is the first step to ensuring the correct computation in this algorithm.

The Steps to This Algorithm

1. Write the larger number on top of the smaller number, making sure to align the numbers by their place value.
2. Multiply the number in the ones place of the bottom number by the number in the ones place of the top number.
3. Continue to multiply the number in the ones place by all the other numbers of the top number.
4. Place a zero in the ones column below the product you just finished. (If you were going to multiply the number in the hundreds place of the bottom number, you would add two zeroes, and so on.)
5. Multiply the number in the tens place of the bottom number by all the numbers in the top number, starting with the ones.
6. Continue to follow the same process for each number in the bottom number.
7. Add to get the final product.

Examples

$74 \times 53 = 3{,}922$

The first step is to multiply the ones digit in the bottom number by the ones digit in the top number. You then multiply the ones digit in the bottom number by all the other digits in the top number. If you need to regroup a number, it is placed at the top of the next place value's number. Make sure to add the number you carried when multiplying the bottom number by that number.

$$\begin{array}{r} {}^{1}\ \ \\ 74 \\ \times 53 \\ \hline 222 \end{array}$$

- multiply $3 \times 4 = 12$
- multiply $3 \times 7 = 21 + 1 = 22$

Then, multiply the next digit in the bottom number by all the digits in the top number. This is where many children make computation errors because they are not realizing the value of the number in the next place.

In this case the number is a 5, but the actual value that is being multiplied by is 50. To help with this error, place a 0 under the ones to help them realize that they need to start lining up their product in the tens place.

```
  2 1
   74    • multiply 5 × 4 = 20
  ×53    • multiply 5 × 7 = 35 + 2 = 37
  ---
  222
+3700 ← place a zero!
-----
```

Finally, add the partial products together to get your final product.

```
  2 1
   74
  ×53
  ---
  222
+3700    • add to get final product
-----
3,922    • 3,922 is the final product
```

If your child is struggling to line up the places in multiplication problems, use graph paper to help them. One strategy I use is to have my students draw a face instead of the zero. For some students, this simple, fun step helps them to remember. My students have named this mysterious face "Bob." Try this, it really works!

```
  2 1
   96
  ×42
  ---
   1
  192
  1
+384☺
-----
4,032
```

$342 \times 54 = 18{,}468$

```
 1
 342    • multiply 4 × 2 = 8
×  54   • multiply 4 × 4 = 16  (place 6; carry the 1 to hundreds)
----
1368    • multiply 4 × 3 = 12 + 1 = 13
```

```
  2 1    • place a zero
  342    • multiply 5 × 2 = 10  (place 0; carry the 1 to tens)
×  54    • multiply 5 × 4 = 20 + 1 = 21  (place 1; carry the 2 to hundreds)
 1368    • multiply 5 × 3 = 15 + 2 = 17
+17100   • add the partial products
------
18,468   • 18,468 is the final answer
```

$342 \times 196 = 67{,}032$

```
  3 2 1 1
      342
     ×196
     2052
    30780   (don't forget to add the zero!)
   +34200
   67,032
```

Practice Problems

3.1 $34 \times 56 =$

3.2 $621 \times 72 =$

3.3 $504 \times 43 =$

3.4 $326 \times 37 =$

3.5 $5{,}654 \times 419 =$

Partial-Products Multiplication

Traditional multiplication can be hard for many children because it involves large numbers. It also uses multiplication and addition as you move through each digit in the problem. Partial-products multiplication has children do all the multiplication steps first and then add the results to get the final answer. In my classroom, many students who struggle to remember when to regroup and misalign the products under the correct factor find this algorithm much easier for them to multiply accurately.

The Steps to This Algorithm

1. Multiply each number in the bottom factor by each number in the top factor. Multiply from left to right, starting with the largest number.
2. Add all the partial products together to get the final product.

Examples

The first step is to multiply each digit in the bottom factor by each digit in the top factor. Be sure to align the partial products when writing them to ensure that the correct place value numbers are being multiplied.

hundreds	tens	ones	
4	2	6	• multiply 8 × 400 = 3200
×		8	• multiply 8 × 20 = 160
			• multiply 8 × 6 = 48

Then, add the partial products from the problem to find the final product.

$$\begin{array}{r} \overset{1}{3}200 \\ 160 \\ +\quad 48 \\ \hline 3{,}408 \end{array}$$

• add the partial products
• 3,408 is the final product

> Some children may struggle with knowing the correct place value to multiply. For example, in the preceding problem, some children might mistakenly multiply 8*2 instead of 8*20. If this is the case, write the value for each digit of the number on the paper before doing the computation.

```
  392   3 = 300
×   8   9 =  90
        2 =   2
        8 =   8
```

Here is another problem with a single digit multiplier.

$692 \times 7 = 4{,}844$

First, multiply each digit in the bottom factor by each digit in the top factor. Then, add the partial products to get your final product.

```
        • multiply 7 × 600 = 4200
  692   • multiply 7 × 90 =   630
×   7   • multiply 7 × 2 =     14
4,844 ← • add the           4,844
          partial products
```

This strategy is just as easy to do when you have a multi-digit multiplier. Follow the same steps as before. You will just have more numbers to multiply and add.

$464 \times 23 = 10{,}672$

```
                                1
         • multiply 20 × 400 = 8000
         • multiply 20 × 60 =  1200
         • multiply 20 × 4 =     80
         • multiply 3 × 400 =  1200
   464   • multiply 3 × 60 =    180
×   23   • multiply 3 × 4 =      12
10,672 ← • add the           10,672
           partial products
```

Here is one more example with a multi-digit multiplier.

$436 \times 73 = 31,828$

	• multiply 70 × 400 =	28000 (carries: 1 1)
	• multiply 70 × 30 =	2100
	• multiply 70 × 6 =	420
	• multiply 3 × 400 =	1200
436	• multiply 3 × 30 =	90
× 73	• multiply 3 × 6 =	18
31,828 ←	• add the partial products	31,828

If your child is having a hard time lining up the partial products, use graph paper to help keep all the places in line. Use a marker to make lines for each place value. Making lines will ensure that all numbers are lined up when adding for the final product.

Practice Problems

3.6 $24 \times 65 =$

3.7 $176 \times 34 =$

3.8 $307 \times 45 =$

3.9 $1{,}753 \times 25 =$

3.10 $4{,}387 \times 398 =$

Traditional Multiplication for Decimals

This algorithm is identical to the steps of traditional multiplication without decimals. If your child is not having a problem multiplying in this way, then this is the way you will want to reinforce computation with decimals. You multiply the two numbers, ignoring the decimals until the end. Once the answer has been computed, you use the original problem to place the decimal point. This is the step where children need to have a firm grasp on number sense.

The Steps to This Algorithm

1. Find the product of the two numbers by following the steps to the traditional multiplication algorithm.
2. After the product has been found, count over the number of places that the decimal is placed in both numbers and then place it that number of places starting from the right in the product. For example, if you move over two places in the first number and two places in the second number you are multiplying, then you will place the decimal four places from the right in the product.

Examples

$4.6 \times 3.5 = 16.1$

First multiply the two numbers, ignoring the decimal points in both numbers. One thing to notice in this problem is that the answer ends in a zero. Keep this zero in the product until you place the decimal point.

$$\begin{array}{r} 46 \\ \times 35 \\ \hline 230 \\ 1380 \\ \hline 1610 \end{array}$$

The product of 46 and 35 is 1,610.

Next, count the number of decimal places in both numbers and place the decimal that number of places in the answer. In this problem, you will place it in two decimal places from the right. The answer can then be 16.10 or 16.1. After you place the decimal, omitting the zero at the end of the answer does not change the value of the number.

4.6×3.5

+1 +1 = 2 places

final answer = 16.10

$24.5 \times 2.4 = 58.8$

First multiply the two numbers, ignoring the decimal points in both numbers. One thing to notice in this problem is that the answer ends in a zero. Keep this zero in the product until you place the decimal point.

$$\begin{array}{r} 245 \\ \times 24 \\ \hline 980 \\ 4900 \\ \hline 5880 \end{array}$$

The product of 245 and 24 is 5,880.

Next, count the number of decimal places in both numbers. Be sure to keep the zero until you place the decimal point. The answer can then be 58.80 or 58.8. After placing the decimal, omitting the zero at the end of the answer does not change the value of the number.

24.5×2.4

$+1 \quad +1 = 2$ places

58.80

final answer = 58.80 **or** 58.8

$60.5 \times 41 = 2{,}480.5$

```
   2
  605
×  41
  605
24200
24805
```

60.5×41

$+1 = 1$ place

final answer = 2,480.5

$0.67 \times 23 = 15.41$

```
 21
 67
×23
201
1340
1541
```

0.67×23

$+2 = 2$ places

final answer = 15.41

Practice Problems

3.11 $1.6 \times 6.2 =$

3.12 $27.6 \times 31 =$

3.13 $51 \times 8.8 =$

3.14 $3.90 \times 4.8 =$

3.15 $4.56 \times 6.4 =$

Partial-Products Box Method

The educational coach at our school introduced this strategy to some of my students who were having a hard time with the organization of partial products. With this strategy, children place the numbers to be multiplied in a box. This gives them visual help while doing this step. The numbers are then added to get the final product. This strategy gave many of my students who were unsuccessful with other algorithms the chance to be successful.

The Steps to This Algorithm

1. Figure out how many boxes you need to make to organize the problem. A problem that multiplies a two-digit number by another two-digit number requires a two-by-two box.
2. Write the value of each of the digits in the first number above the top two boxes. For example, if the first number is 25, write 25 = 20 + 5. Do the same thing for the second number to the left of the left-most boxes. For example, if the second number is 82, write 82 = 80 + 2.
3. Multiply the numbers that meet for each box. You are essentially doing the same thing as partial products, but in a concrete, organized manner.
4. After you finish multiplying all the boxes, add all the products together to get the final product. I have my students break this into two steps, adding the columns first and then adding those sums together to get the final answer.

Examples

$54 \times 26 = 1{,}404$

	50 ↓	4 ↓
20 →	1,000	80
6 →	300	24
	1,300	104

$$\begin{array}{r} 1{,}300 \\ +\ 104 \\ \hline 1{,}404 \end{array}$$

$456 \times 57 = 25{,}992$

	400	50	6
50	20,000	2,500	300
7	2,800	350	42
	22,800	2,850	342

$$\begin{array}{r} \overset{1}{2}2{,}800 \\ 2{,}850 \\ +\ 342 \\ \hline 25{,}992 \end{array}$$

703 × 49 = 34,447

	700	0	3
40	28,000	0	120
9	6,300	0	27
	34,300	0	147

```
  34,300
+    147
  34,447
```

Practice Problems

3.16 59 × 92 =

3.17 204 × 45 =

3.18 517 × 639 =

3.19 3,295 × 439 =

3.20 2,659 × 85 =

Lattice Multiplication

Lattice multiplication is an algorithm used by a large majority of my students. This algorithm allows students to multiply large numbers with little error because it solves by using basic math facts. The problem is solved by drawing a large grid and using simple multiplication facts for each step in the problem. This is the perfect way to solve larger multiplication problems for children who struggle with all other methods. In most cases, children are completely successful with this method once they get the hang of it.

The Steps to This Algorithm

1. Draw a grid, called a lattice, which has the same number of columns as there are digits in the multiplicand (first number) and the same number of rows as there are digits in the multiplier (second number).
2. Draw a diagonal line from the upper-right corner of each box in the grid through the lower-left corner and a bit beyond.
3. Write the multiplicand on the top of the lattice and the multiplier to the right of the lattice.
4. Multiply the first number at the top of the column by the number to the right of the row. Then write the product in the square with the tens above the diagonal line and the ones below. If the product does not have a tens digit, place a zero in the space.
5. Add the diagonals and carry to the next diagonal if needed. Start at the box in the lower-right corner. The final answer is given at the bottom of the diagonals when you read the number from left to right.

Examples

$54 \times 7 = 378$

First, make a lattice that is two columns by one row. Draw a diagonal line from the upper-right corner of each box in the grid to the lower-left corner.

Be surc you extend the lines beyond the bottom of the boxes to keep the numbers separated.

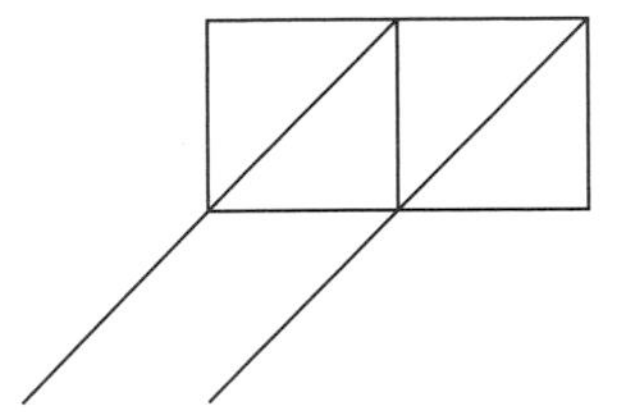

Next, write the multiplicand on the top of the lattice and the multiplier on the right.

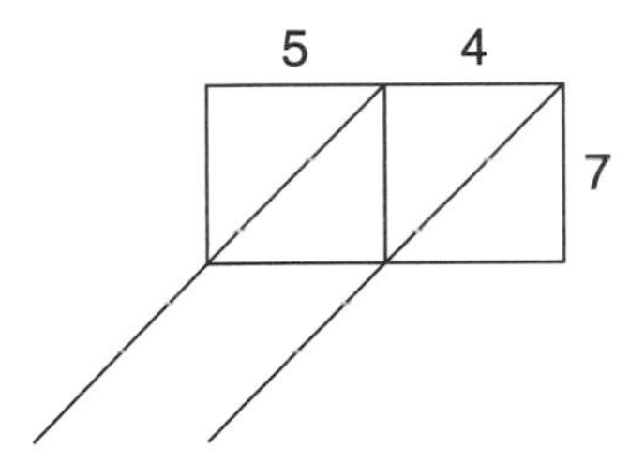

Multiply thc number at the top by the number at the right for each box. Write the product in the boxes with the tens above the diagonal line and the ones below.

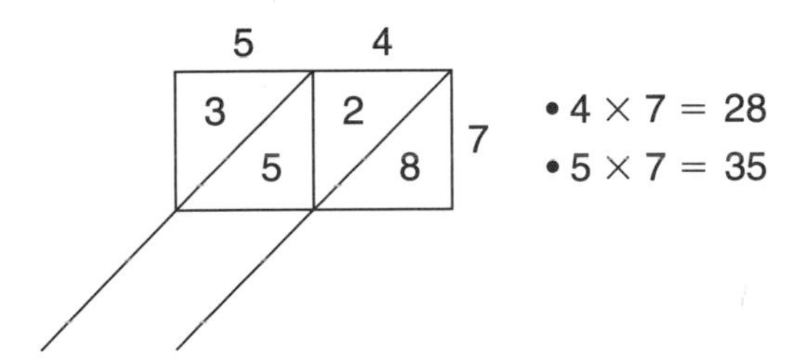

Finally, add the numbers in each diagonal from right to left. The final answer is read from left to right at the bottom of the diagonals.

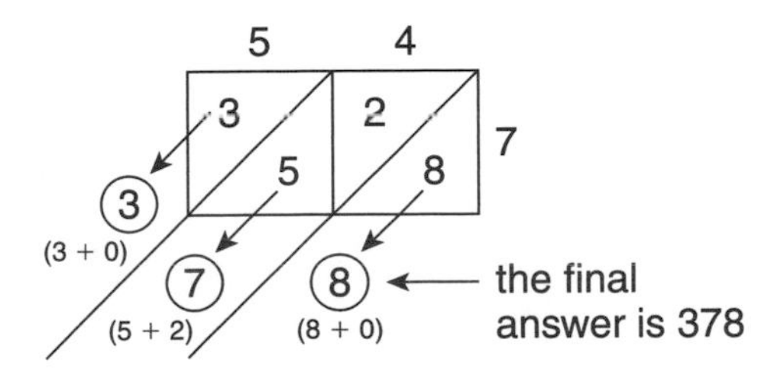

$82 \times 34 = 2{,}788$

First, make a 2-by-2 lattice and draw a diagonal line from the upper-right corner of each box in the lattice to its lower-left corner.

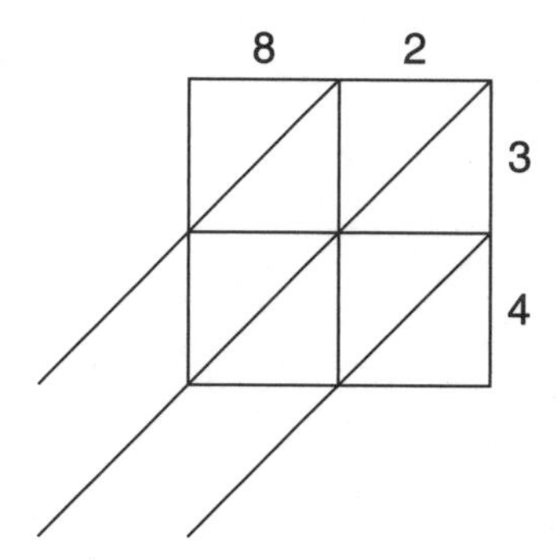

Next, multiply the number at the top by the number at the right for each box. Write the product in the boxes with the tens above the diagonal line and the ones below.

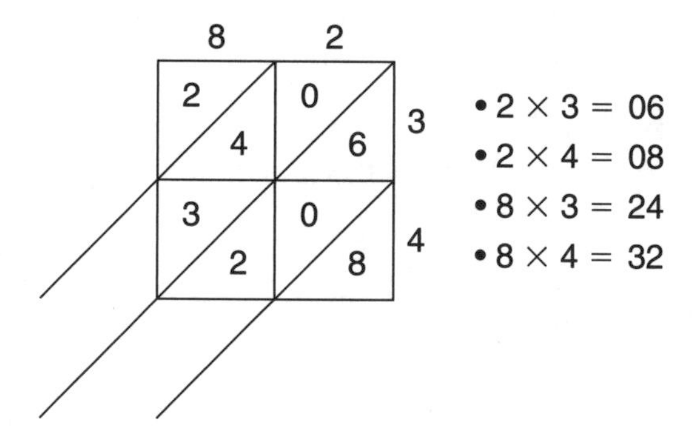

Finally, add the numbers in each diagonal from right to left. The final answer can be read from left to right at the bottom of the diagonals.

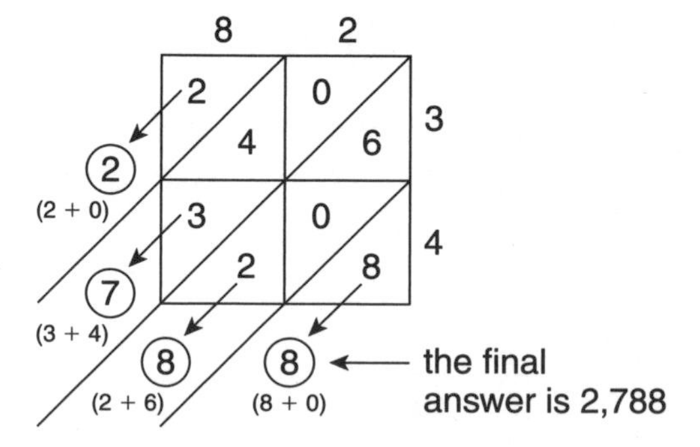

$342 \times 76 = 25{,}992$

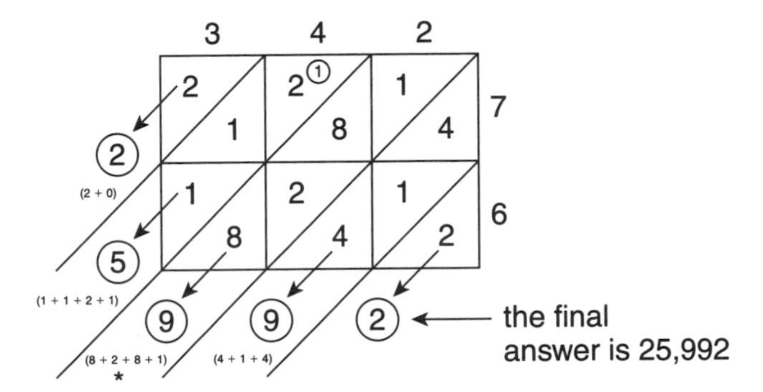

* In this problem, when adding the third diagonal, you need to carry the 1 to the next column because the answer totals 19.

$276 \times 342 = 94{,}392$

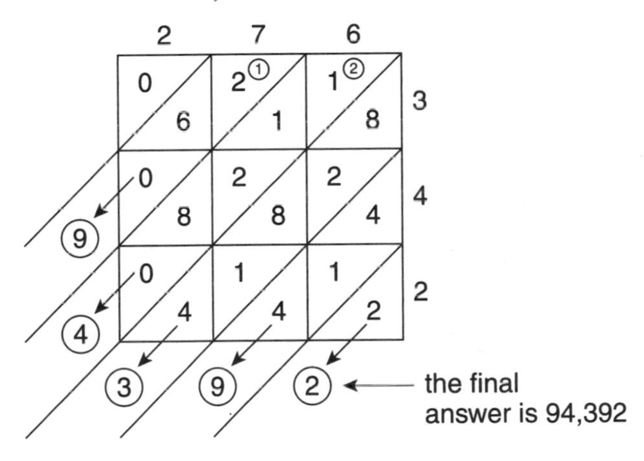

Practice Problems

3.21 $74 \times 23 =$

3.22 $908 \times 41 =$

3.23 $823 \times 736 =$

3.24 $2,495 \times 628 =$

3.25 $3,734 \times 78 =$

Lattice Multiplication for Decimals

This algorithm is identical to the steps of lattice multiplication without decimals. If your child is not having a problem multiplying in this way, then this is the way you will want to reinforce computation with decimals. The main difference is the steps that are taken to place the decimal in the correct place in the final answer. This is done by finding the intersection of the decimal points along the horizontal and vertical lines.

The Steps to This Algorithm

1. Make the lattice and follow the steps to lattice multiplication with whole numbers.
2. Find the intersection of the decimal points by running a line along the horizontal and vertical lines where the decimals are in each number.
3. Finally, slide it down along the diagonal where the two points meet.

Examples

2.7 × 5.6 = 15.12

First, multiply following the steps of the lattice multiplication algorithm.

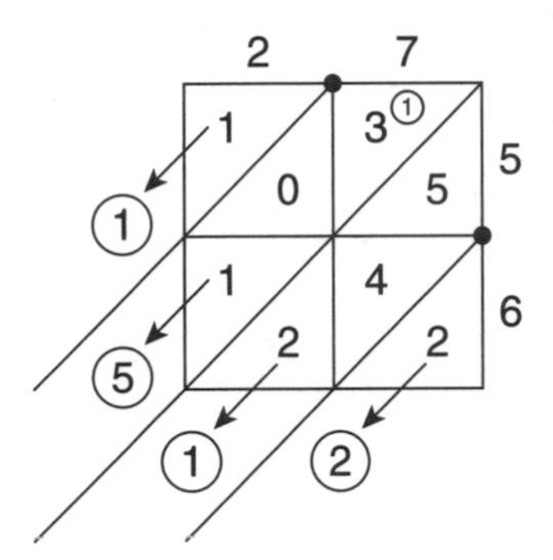

Then, find the intersection of the decimal points by running a line along the horizontal and vertical lines where the decimals are in each number. Finally, slide it down along the diagonal where the two points meet and place it in the answer.

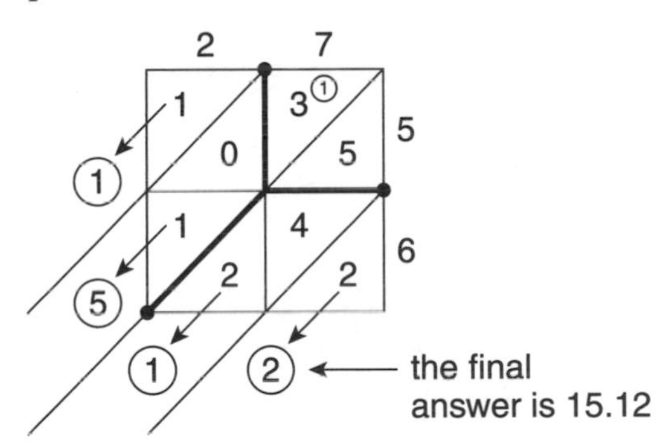

3.45 × 4.2 = 14.49

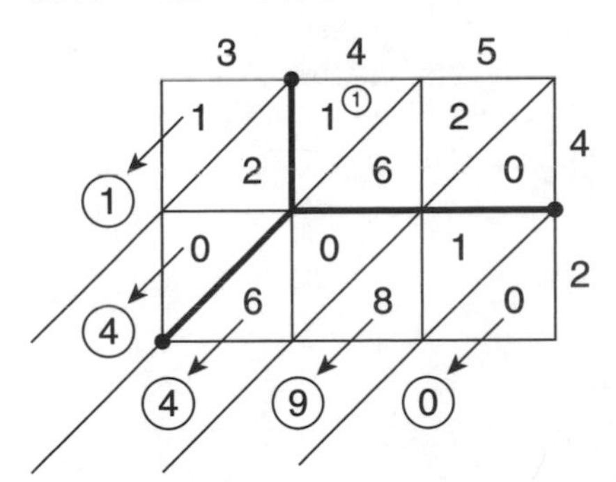

$23 \times 6.5 = 149.5$

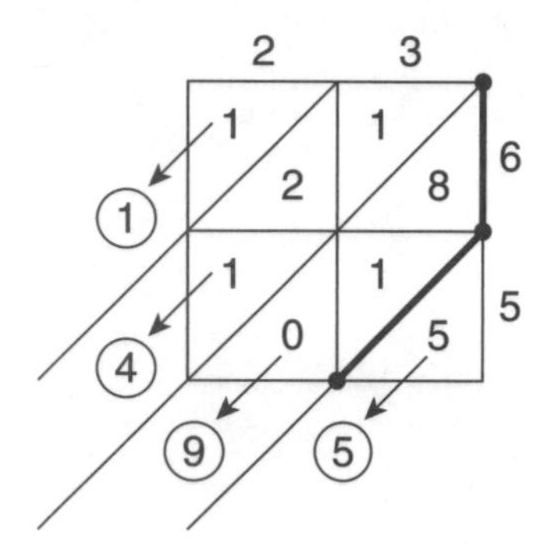

- in this problem you are multiplying by a whole number
- look closely at how to place the decimal by using the diagonal

$8.67 \times 5.9 = 51.153$

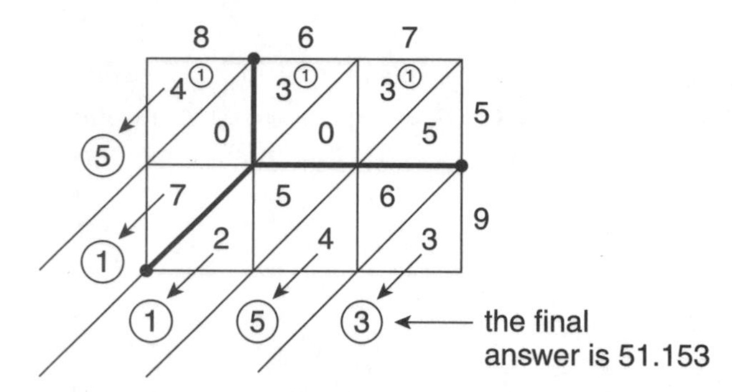

Many students struggle to find the placement of the decimal because they can't find the intersection. If this is an issue, multiply using lattice and then use the method for finding the decimal in the traditional multiplication algorithm. Once the product has been found using lattice, count how many total places there are after the decimal point in the multiplicand and multiplier. Put the decimal point that number of places over starting from the right in the product. For example, if you move over two places in the first number and two places in the second number you are multiplying, then you will place the decimal four places starting from the right in the product.

Practice Problems

3.26 $7.1 \times 3.9 =$

3.27 $4.22 \times 3.4 =$

3.28 $67 \times 8.9 =$

3.29 $456 \times 2.5 =$

3.30 $70.6 \times 4.4 =$

Chapter 4

Division Algorithms

Traditional Long Division

Most likely, traditional long division is the algorithm with which you are most familiar. You're probably thinking, why would anyone divide any other way? In traditional division, the problem is broken into smaller steps that include division, multiplication, and subtraction. The first number, called the dividend, is divided by the second number, called the divisor. The resulting answer is called the quotient.

You create the symbol separating the divisor from the dividend by placing a right parenthesis with an attached vinculum (straight line) drawn to the right. This symbol does not have a specific name, so I will refer to it as the long division symbol.

$$\text{divisor}\overline{)\text{dividend}}$$

Many of my students struggled dividing in this way until I introduced them to a mnemonic saying that made the steps clear and understandable. I don't know who came up with this saying, but whoever it was is a genius! In this mnemonic saying, the names of the members of a family represent a step in the division process:

- **Dad:** Divide
- **Mom:** Multiply
- **Sister:** Subtract
- **Brother:** Bring down
- **Rover:** Repeat/remainder

Many of the students in my classroom write "DMSBR" at the top of the page to help them remember the steps. This has helped so many of my students who were not successful with division in the past to divide with ease. If you start using this from the beginning, the steps will become automatic when your child learns division.

The Steps to This Algorithm

1. Write the dividend (the first number) under the long division symbol and the divisor (the second number) on the outside to the left.
2. Divide (Dad) to see how many times the divisor can go into the first digit or first two digits of the dividend without going over. Place that digit in the quotient, above the number that you are dividing into.
3. Multiply (Mom) the number you placed in the quotient by the divisor. Place the product under the number in the dividend.
4. Subtract (Sister) the product from step 3 from the number in the dividend.
5. Bring down (Brother) the next number in the dividend, placing it alongside the number that resulted from the subtraction.
6. Repeat (Rover) until there are no longer any numbers left to bring down. The number that remains becomes the remainder.

Examples

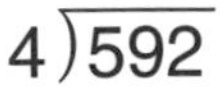

$4\overline{)592}$ $592 \div 4 = 148$

Begin by writing the dividend inside the division symbol and the divisor on the outside.

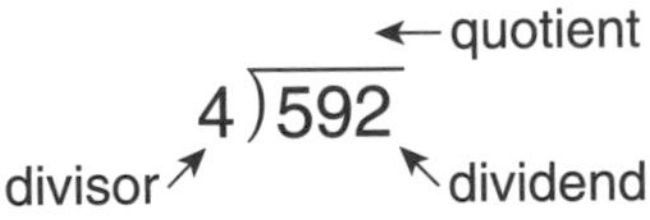

←quotient
$4\overline{)592}$
divisor↗ ↖dividend

The first step is to see how many times the divisor will divide (Dad) into the first number in the dividend without going over. How many times will 4 divide (Dad) into 5? It will divide in one time without going over. Place a 1 in the quotient directly above the 5 in the dividend.

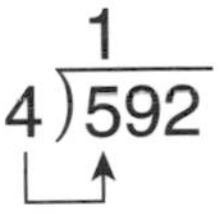

$$\begin{array}{r} 1 \\ 4\overline{)592} \end{array}$$

(Dad – Divide)
$5 \div 4 = 1$

Next, multiply (Mom) the number you just placed in the quotient by the divisor. Put the product under the first number in your dividend.

$$\begin{array}{r} 1 \\ 4\overline{)592} \\ \underline{4} \end{array}$$

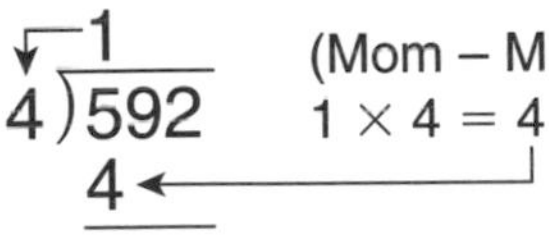

(Mom – Multiply)
$1 \times 4 = 4$

Then, subtract (Sister) the product you just wrote down from the number above it (the first number in your dividend).

$$\begin{array}{r} 1 \\ 4\overline{)592} \\ \underline{-4} \\ 1 \end{array}$$

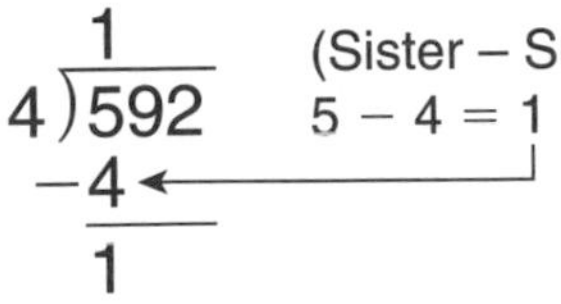

(Sister – Subtract)
$5 - 4 = 1$

Finally, bring down (Brother) the next number in the dividend.

$$\begin{array}{r} 1 \\ 4\overline{)592} \\ \underline{-4\downarrow} \\ 19 \end{array}$$

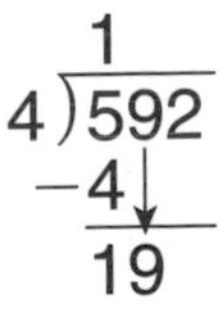

(Brother – Bring Down)
bring down the 9

Now, repeat (Rover) the same steps in order. First, see how many times 4 will divide (Dad) into 19. It will go in four times without going over. Place a 4 in the quotient, directly above the 9 in the dividend.

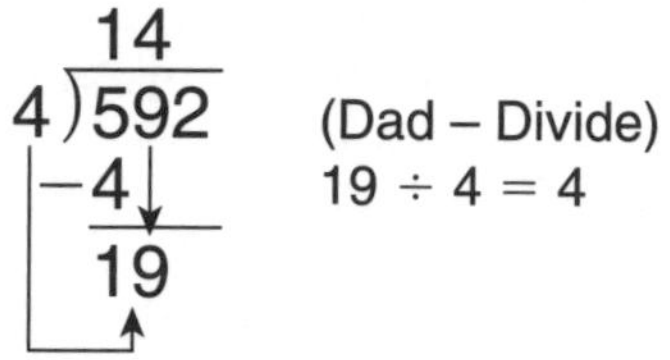

```
   14
4)592     (Dad – Divide)
 -4↓      19 ÷ 4 = 4
  19
```

Next, multiply (Mom) the number you just placed in the quotient by the divisor. Put the product under the numbers below the dividend.

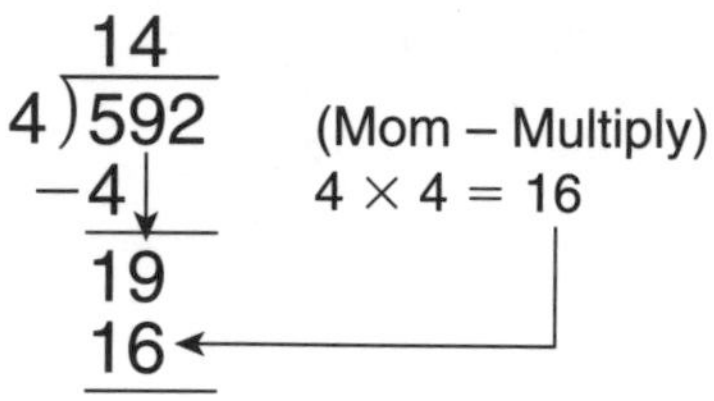

```
   14
4)592     (Mom – Multiply)
 -4↓      4 × 4 = 16
  19
  16
```

Then, subtract (Sister) the number on the bottom from the top number.

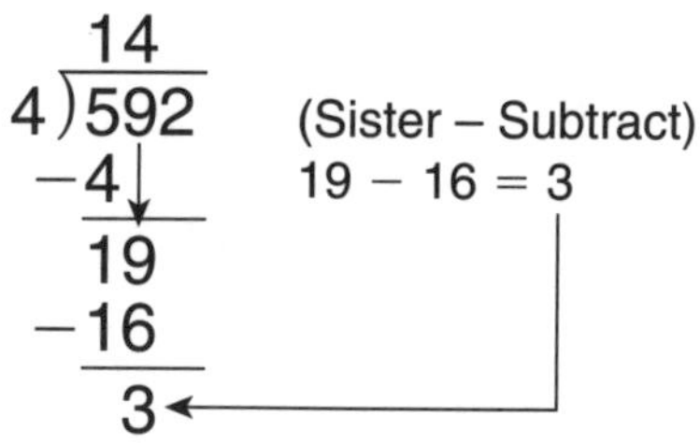

```
   14
4)592     (Sister – Subtract)
 -4↓      19 − 16 = 3
  19
 -16
   3
```

Finally, bring down (Brother) the next number in the dividend.

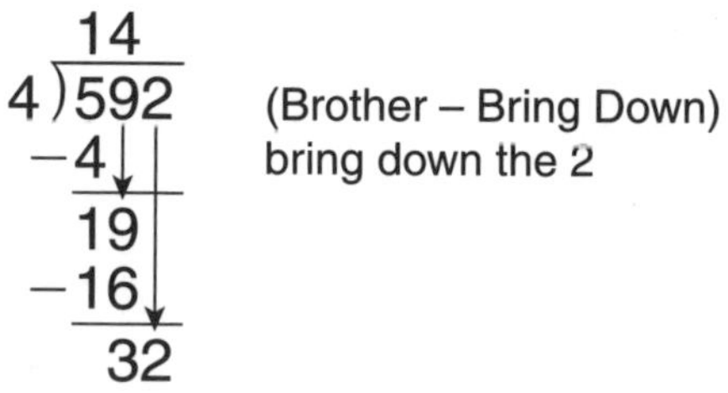

```
   14
4)592     (Brother – Bring Down)
 -4↓      bring down the 2
  19
 -16↓
   32
```

Now, repeat (Rover) the same steps in order. This is the last series of the steps until you get to the quotient. First, estimate how many times 4 will divide (Dad) into 32. It will go in eight times without going over. Place an 8 in the quotient directly above the 2 in the dividend.

$$\begin{array}{r} 148 \\ 4\overline{)592} \\ -4 \\ \hline 19 \\ -16 \\ \hline 32 \end{array}$$

(Dad – Divide)
$32 \div 4 = 8$

Next, multiply (Mom) the number in your quotient by the divisor. Place the product under the next number in your dividend.

$$\begin{array}{r} 148 \\ 4\overline{)592} \\ -4 \\ \hline 19 \\ -16 \\ \hline 32 \\ 32 \\ \hline \end{array}$$

(Mom – Multiply)
$8 \times 4 = 32$

Then, subtract (Sister) the number on the bottom from the number above it.

$$\begin{array}{r} 148 \\ 4\overline{)592} \\ -4 \\ \hline 19 \\ -16 \\ \hline 32 \\ -32 \\ \hline 0 \end{array}$$

(Sister – Subtract)
$32 - 32 = 0$

There are no other numbers to bring down (Brother). If there is a number left after you subtract, that number becomes the remainder (Rover).

```
   148 r0
4)592
 -4
  19
 -16
   32
  -32
    0   (Rover – Remainder)
```

592 ÷ 4 = 148 r0

Now for the next example:

9)237   237 ÷ 9 = 26 r3 or $26\frac{3}{9} = \frac{1}{3}$

The first step is to see how many times the divisor will divide (Dad) into the first number in the dividend without going over. In this problem, the 9 will not divide into the 2, so move over one more place in the dividend to see how many times 9 will divide into 23. The answer is 2; place the 2 in the quotient directly above the 3 in the dividend.

Then, follow the same steps in order:

```
  x2
9)237
 -18
   57
```

- Dad – Divide 23 ÷ 9 = 2
- Mom – Multiply 2 × 9 = 18
- Sister – Subtract 23 − 18 = 5
- Brother – Bring Down the 7

Now, repeat (Rover) the same steps in order.

```
  x26 r3
9)237
 -18
   57
  -54
    3
```

- Dad – Divide 57 ÷ 9 = 6
- Mom – Multiply 6 × 9 = 54
- Sister – Subtract 57 − 54 = 3
- There is not a number to bring down
- Rover – Remainder

The remainder in the quotient can be written as a remainder or as a fraction. This will depend on how you are asked to report the remainder. If you are reporting the remainder as a fraction, the remainder becomes the numerator of the fraction (the top number) and the divisor becomes the denominator of the fraction (the bottom number). (Note that you'll find information on reducing and simplifying fractions in Chapter 5, "Fraction Concepts.")

$237 \div 9 = 26$ r3 or $26\frac{3}{9}$ ($26\frac{1}{3}$ reduced to simplest form)

> Notice in this problem that an X was placed over the 2 because 9 would not divide into 2. Placing an X here will help your child to always write the number over the correct place in the dividend. This is a common error with children; placing the X helps them to visualize that step.

The steps are the same whether there is a single digit or multi-digit divisor:

$15\overline{)497}$ $497 \div 15 = 33$ r2 or $33\frac{2}{15}$

The first step is to see how many times the divisor will divide (Dad) into the dividend without going over. In this problem, the 15 will not divide into the 4, so move over one more place and see how many times 15 will divide into 49. The answer is 3; place a 3 in the quotient directly above the 9 in the dividend.

Then, follow the same steps in order:

```
   x3
15)497
  -45↓
    47
```

- Dad – Divide $49 \div 15 = 3$
- Mom – Multiply $3 \times 15 = 45$
- Sister – Subtract $49 - 45 = 4$
- Brother – Bring Down the 7
- Rover – Repeat

Now, repeat (Rover) the same steps in order.

```
     xx33 r2
15 ) 497
    -45↓
      47
    - 45
       2
```

- Dad – Divide 47 ÷ 15 = 3
- Mom – Multiply 3 × 15 = 45
- Sister – Subtract 47 − 45 = 2
- There is not a number to bring down
- Rover – Remainder

$497 \div 15 = 33 \text{ r2 or } 33\frac{2}{15}$

Here's another example:

```
34 ) 1942
```

1,942 ÷ 34 =

The first step is to see how many times the divisor will divide (Dad) into the dividend without going over. You will notice in this problem that the 34 will not divide into the 1 or 19, so you must move over one more place to see how many times 34 will divide into 194. The answer is 5; place a 5 in the quotient directly above the 4 in the dividend.

Then, follow the same steps in order:

```
      xx5
34 ) 1942
    -170↓
      242
```

- Dad – Divide 194 ÷ 34 = 5
- Mom – Multiply 5 × 34 = 170
- Sister – Subtract 194 − 170 = 24
- Brother – Bring Down the 2
- Rover – Repeat

Now, repeat (Rover) the same steps in order.

```
      xx57 r4
34 ) 1942
    -170
      242
    - 238
        4
```

- Dad – Divide 242 ÷ 34 = 7
- Mom – Multiply 7 × 34 = 238
- Sister – Subtract 242 − 238 = 4
- There is not a number to bring down
- Rover – Remainder

$1{,}942 \div 34 = 57 \text{ r}4$ or $57\frac{4}{34}$ ($57\frac{2}{17}$ reduced to simplest form)

> Another strategy that helps my students is to underline the numbers that you will be dividing into on the first step. They do this as they place the X in the quotient. This helps them to focus on only the numbers they will be using to divide in the first step.
>
> $$9\overline{)\underline{23}9} \quad \text{(quotient: X)}$$
>
> underline the 23 to put focus on the number being used

This last example explains what to do when you cannot divide the divisor into the number that was carried down.

$7\overline{)4923}$ $4923 \div 7 = 703 \text{ r}2$ or $703\frac{2}{7}$

Follow the same steps in order:

```
   x7
7)4923
 -49↓
   02
```

- Dad – Divide $49 \div 7 = 7$
- Mom – Multiply $7 \times 7 = 49$
- Sister – Subtract $49 - 49 = 0$
- Bring Down the 2
- Rover – Repeat

Now, repeat (Rover) the same steps in order. You will quickly see that 7 cannot divide into 2. In this case, you place a 0 in your quotient because you cannot divide 7 into 2. Then, bring down the next number.

```
   x70
7)4923
 -49↓↓
   023
```

Now, repeat (Rover) the same steps in order.

```
  x703 r2
7)4923
 -49↓↓
   023
 -  21
     2
```

- Dad – Divide $23 \div 7 = 3$
- Mom – Multiply $3 \times 7 = 21$
- Sister – Subtract $23 - 21 = 2$
- There is nothing to bring down
- Rover – Remainder

$4923 \div 7 = 703$ r2 or $703\frac{2}{7}$

Practice Problems

4.1 $852 \div 6 =$

4.2 $1,452 \div 8 =$

4.3 $5,763 \div 12 =$

4.4 $976 \div 24 =$

4.5 $6,485 \div 45 =$

Traditional Long Division with Decimals

This algorithm is identical to the steps of traditional division without decimals. This would be the method that I would suggest using if dividing with decimals. If your child is not having a problem with dividing in this way, then this is the way you will want to reinforce computation with decimals. You divide the two numbers by following the same steps as for traditional division. Once the answer has been computed, you carry the decimal into the quotient.

The Steps to This Algorithm

1. Write the dividend inside the long division symbol and the divisor on the outside.
2. Carry the decimal up into the quotient if there is only a decimal in the dividend. If there is a decimal in the divisor, follow the steps in the examples that follow before dividing.
3. Divide following the steps for traditional division: divide, multiply, subtract, bring down, repeat.

Examples

In the first two problems, there is a decimal only in the dividend. In this case, the first step after writing the problem is to carry the decimal up immediately to the quotient before doing any computation. You will then divide using the traditional algorithm and ignore the decimal point.

$7\overline{)64.61}$ $64.61 \div 7 = 9.23$

First, immediately bring the decimal into the quotient. Lining up numbers as you divide is crucial.

$$\begin{array}{r} . \\ 7\overline{)64.61} \end{array}$$

Next, divide following the steps for traditional division: divide, multiply, subtract, bring down, and repeat. Do this until you cannot divide any longer, or until you have divided to the designated place value beyond the decimal point.

> In many instances, a decimal will be able to divide to an indefinite number of places. In such cases, the instructions should say to divide the decimal to the hundredths, thousandths, or another place. The problem should be divided until that place or until you cannot divide any longer.

$$\begin{array}{r} \text{x}9. \\ 7\overline{)64.61} \\ -63 \\ \hline 16 \end{array}$$

- Divide $64 \div 7 = 9$
- Multiply $9 \times 7 = 63$
- Subtract $64 - 63 = 1$
- Bring Down the 6
- Repeat

Repeat the same steps in order.

$$\begin{array}{r} 9.2 \\ 7\overline{)64.61} \\ -63 \\ \hline 16 \\ -14 \\ \hline 21 \end{array}$$

- Divide $16 \div 7 = 2$
- Multiply $2 \times 7 = 14$
- Subtract $16 - 14 = 2$
- Bring Down the 1
- Repeat

Continue to repeat the same steps in order.

$$\begin{array}{r} 9.23 \\ 7\overline{)64.61} \\ -63 \\ \hline 16 \\ -14 \\ \hline 21 \\ -21 \\ \hline 0 \end{array}$$

- Divide $21 \div 7 = 3$
- Multiply $3 \times 7 = 21$
- Subtract $21 - 21 = 0$
- The difference is zero, so you cannot divide any longer

$64.61 \div 7 = 9.23$

Following is another example:

```
      .
12)991.2
```

991.2 ÷ 12 = 82.6

Immediately carry the decimal to the quotient and divide using the traditional algorithm, ignoring the decimal.

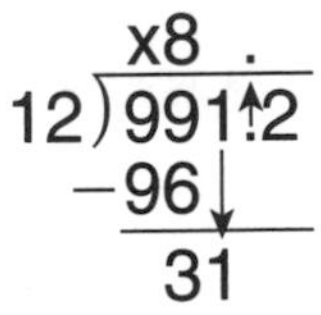

```
    x8 .
12)991.2
  −96
    31
```

Now, repeat the same steps in order.

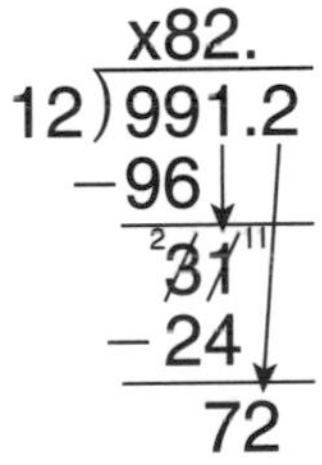

```
    x82.
12)991.2
  −96
    31
   −24
     72
```

Continue to repeat the steps in order.

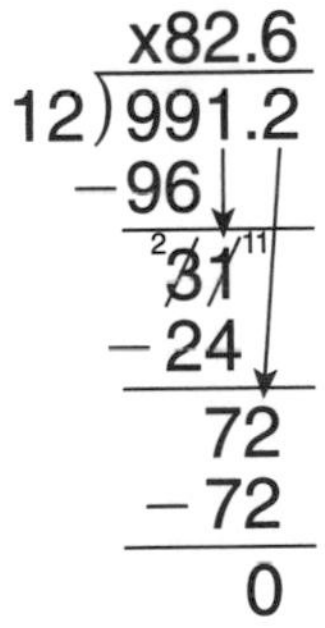

```
    x82.6
12)991.2
  −96
    31
   −24
     72
    −72
      0
```

991.2 ÷ 12 = 82.6

The next two problems are examples that have decimals in the divisor *and* the dividend. When this is the case, the decimal needs to be removed from the divisor prior to dividing. This changes the divisor from a decimal to a whole number. Then, division follows the same steps in the algorithm.

$0.9\overline{)4.68}$ $4.68 \div 0.9 = 5.2$

The first step is to remove the decimal from the divisor. Count the number of places you need to move the decimal to make it a whole number. In this case, you need to move the decimal one place to the right.

The next step is to move the decimal in the dividend the same number of places. You are essentially multiplying both numbers by a power of 10 when you move the decimal points. Because the decimal in this problem was moved one place, you are multiplying both numbers by 10.

$0.9\overline{)4.68} \longrightarrow 9\overline{)46.8}$

move decimal one place to the right to make the divisor 9

new problem once decimals are moved

$$
\begin{array}{r}
x5.2 \\
9\overline{)46.8} \\
-45\downarrow \\
\hline
18 \\
-18 \\
\hline
0
\end{array}
$$

- carry the decimal up to the quotient
- divide

$4.68 \div 0.9 = 5.2$

Here's another example:

$0.75\overline{)6.825}$ $6.825 \div 0.75 =$

The first step is to remove the decimal from the divisor. In this case, the decimal needs to be moved two places to the right to make the divisor 75.

The next step is to move the decimal in the dividend two places. Because you are moving the decimal two places, you are essentially multiplying both numbers by 100.

$0.75\overline{)6.825} \longrightarrow 75\overline{)682.5}$

move decimal two places to the right to make the divisor 75

new problem once the decimals are moved

$$\begin{array}{r} xx9.1 \\ 75\overline{)682.5} \\ -675 \\ \hline 75 \\ -75 \\ \hline 0 \end{array}$$

- carry the decimal up to the quotient
- divide

6.825 ÷ 0.75 = 9.1

Practice Problems

4.6 3.42 ÷ 0.9 =

4.7 5.04 ÷ 0.6 =

4.8 5.88 ÷ 1.2 =

4.9 $0.312 \div 0.6 =$

4.10 $2.196 \div 0.3 =$

Partial quotients can also be used to divide with decimals. This is easy to do when the numbers divide equally, but when you have numbers that do not divide equally, it can become difficult to understand how to create more places after the decimal point in the dividend to extend the number. I have found that teaching my students traditional division is the better way to divide with decimals. The setup of traditional division makes it easier to extend numbers beyond the decimal point in the dividend.

Partial-Quotients Division

This is the algorithm that most of my students use to divide. This approach breaks the problem into smaller division steps to get to the final quotient. The partial quotients are then added to get the answer. The one problem that many of my students have is that they do not make an estimate that is high enough when solving the smaller problems. This causes them to have many more steps than are necessary and leads to a higher chance of errors in computation. However, this can be a quick algorithm for students to feel comfortable using because they can use numbers that are easy for them to divide. Even though it may take more steps to get to the final answer, there are many different ways to divide to get the correct quotient.

The Steps to This Algorithm

1. Begin by writing the dividend under the long division symbol and the divisor on the outside. Then draw a long line down from the end of the long division symbol to look like a "hangman" game.

2. Divide the divisor into the dividend by estimating how many groups of the divisor are found in the dividend.
3. Continue to make partial quotients and subtract them from the remaining numbers.
4. When you cannot take any more groups from the number, add the partial quotients. The number left becomes your remainder.

Examples

$3\overline{)294}$ (dividend) (divisor)
294 ÷ 3 = 98

Begin by writing the dividend inside the long division symbol and the divisor on the outside. Then draw a long line down from the end of the division symbol to look like a "hangman" game.

$3\overline{)294}\,|$

Estimate how many times the divisor will divide into the dividend without going over. In this case, estimate how many times 3 will divide into 294. It will divide in about 90 times. Place the 90 on the outside of the hangman line and 270 (3 multiplied by 90) under the dividend. Then, subtract the bottom number you placed from the dividend: 294 – 270 = 24.

$$\begin{array}{r|l} 3\overline{)294} & \\ -270 & 90 \\ \hline 24 & \end{array}$$

Next, estimate how many times 3 will divide into 24. It will divide in eight times exactly. Place an 8 on the outside of the hangman line and 24 under the dividend. Then, subtract the bottom number you placed from the dividend: 24 – 24 = 0. Because you ended with a zero, you will not have a remainder in your quotient.

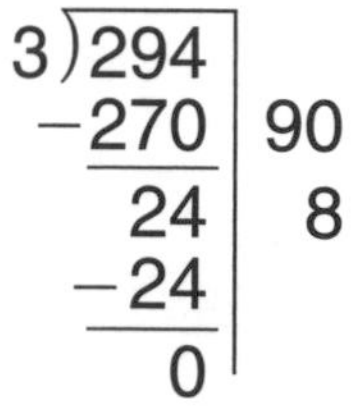

```
3)294 |
 -270 | 90
   24 |  8
  -24 |
    0 |
```

Finally, add the partial quotients to get your final answer: 90 + 8 = 98.

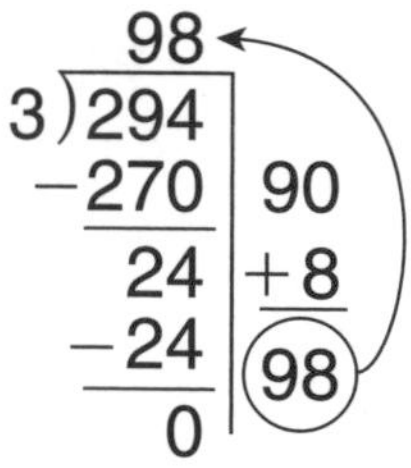

```
   98
3)294 |
 -270 | 90
   24 | +8
  -24 | (98)
    0 |
```

294 ÷ 3 = 98

Estimating can be difficult for some children. If your child struggles with estimating, before making the estimate, underline the first two numbers in the dividend and use basic multiplication facts to make the estimate. This will help by using a smaller number. Then place zeroes under the other places and in the partial quotient. In this problem, decide how many groups of 6 you can make from 38. You can make six, so 6 × 6 = 36. Then add two zeroes to make your first partial quotient 600 × 6 = 3,600.

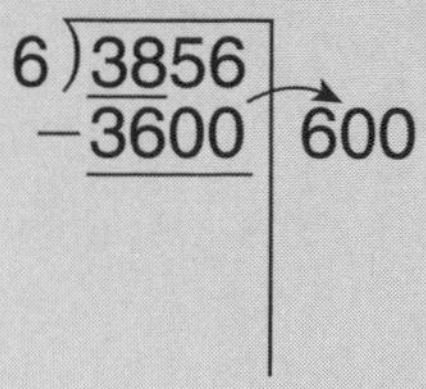

```
6)3856 |
 -3600 | 600
       |
```

$7\overline{)696}$ $696 \div 7 = 99 \text{ r3}$ or $99\frac{3}{7}$

Begin by writing the dividend inside the long division symbol and the divisor on the outside. Then draw a long line down from the end of the division symbol to look like a "hangman" game.

```
7)696 |
      |
      |
      |
```

The first step is to see how many times the divisor will divide into the dividend without going over. Estimate how many times 7 will divide into 696. It will divide in about 90 times. Place a 90 on the outside of the hangman line and a 630 under the dividend. Then, subtract the bottom number you placed from the dividend: 696 – 630 = 66.

```
7)696 |
 -630 | 90
 ----
   66 |
      |
```

Next, estimate how many times 7 will divide into 66. It will divide in nine times. Place a 9 on the outside of the hangman line and a 63 under the dividend. Then, subtract the bottom number you placed from the dividend: 66 – 63 = 3. There are not any more groups of 7 that can be subtracted, so the 3 becomes the remainder.

```
7)696 |
 -630 | 90
 ----
   66 |  9
  -63 |
  ---
    3 |
```

Finally, add the partial quotients to get your final answer: $90 + 9 = 99$.

$696 \div 7 = 99\text{ r}3$ or $99\frac{3}{7}$

```
      99 r3
  7)696  |
   -630  | 90
     66  | +9
    -63  | (99)
      3  |
```

Here's another example:

$24\overline{)4972}$ $4{,}972 \div 24 = 207\text{ r}4$ or $207\frac{1}{6}$

How many times will 24 divide into 4,972? It will divide in about 200 times. Place a 200 on the outside of the hangman line and a 4,800 under the dividend. Then, subtract the bottom number you placed from the dividend: 4,972 – 4,800 = 172.

```
24)4972  |
  -4800  | 200
    172  |
```

Next, estimate how many times 24 will divide into 172. You want to use numbers that are easy to subtract. The easiest would be to make 24 groups of 10, but that is too high. So, five groups of 24 is 120. Place a 5 on the outside and a 120 under the dividend. Then, subtract the two numbers: 172 – 120 = 52.

```
24)4972  |
  -4800  | 200
    172  |   5
   -120  |
     52  |
```

After you subtract, you can still take two more groups of 24 from 52. Place a 2 on the outside and a 48 under the dividend. Subtract the two numbers: 52 – 48 = 4. This is your remainder. Add the partial quotients to get your final answer: 200 + 5 + 2 = 207.

```
      207 r4
24 ) 4972
    -4800 | 200
      172 |   5
     -120 | + 2
       52 | 207
     - 48 |
        4 |
```

$4{,}972 \div 24 = 207$ r4 or $207\frac{4}{24}$ ($207\frac{1}{6}$ reduced to simplest form)

54) 3765 3,765 ÷ 54 = 69 r39 or $69\frac{39}{54}$

Notice that in the following example, I broke the partial quotients into multiples of 10. This is the easiest method for students. It may take a few steps longer, but it will get to the final answer.

```
      69 r39
54 ) 3765
    -2700 | 50
     1065 | 10
     -540 |  9
      525 |
     -486 |
       39 |
```

$3{,}765 \div 54 = 69$ r39 or $69\frac{39}{54}$

Practice Problems

4.11 $752 \div 5 =$

4.12 $2{,}921 \div 6 =$

4.13 $3{,}037 \div 8 =$

4.14 $4{,}287 \div 15 =$

4.15 $7{,}348 \div 35 =$

Chapter 5

Fraction Concepts

Fractions are one of the hardest concepts for children to grasp and master. There are so many names for fractions: simplest form, improper, mixed number, equivalent, and so on. Worse, there are many steps and rules for all these different types of fractions. A child needs to have a clear grasp on fraction concepts before doing any computation or conversion. If your child is struggling to understand fractions, read on to find out how you can help.

Understanding Fraction Parts

Simply put, a fraction is a part of a whole. The top number is called the numerator and the bottom number is called the denominator. A saying that helps my students remember the parts of a fraction is, "The nurse comes in before the doctor." This helps them remember that the numerator is on top and the denominator is on the bottom.

$$\frac{1}{2} = \frac{\text{numerator (nurse)}}{\text{denominator (doctor)}}$$

Fractions are easier to learn if they are shown to students visually with pictures. One way to do this is to show fractions through something that children know well: pizza or pies. Break the pizza or pie into pieces of the fraction. The bottom number of the fraction represents the total number of pieces, and the top number represents the number of pieces that will be shaded, or the part that is eaten.

$$\frac{4}{6} = \frac{\text{shaded pieces}}{\text{total pieces}} \qquad \frac{1}{3} =$$

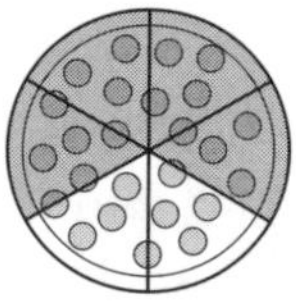

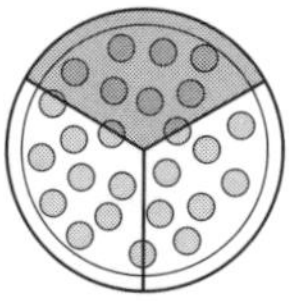

Equivalent fractions are fractions that have different numerators and denominators, but are still equal. Drawing bars will help children to visualize equivalent fractions. This picture shows that the fraction $\frac{1}{3}$ is equivalent to $\frac{3}{9}$. The following fractions are equivalent because the parts that are shaded are the same size. The difference is that the second fraction was broken into nine total parts, with three shaded. There are many ways to represent equivalent fractions. This is explained later in the chapter.

$$\frac{1}{3}$$

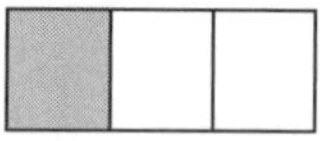

$$\frac{3}{9}$$

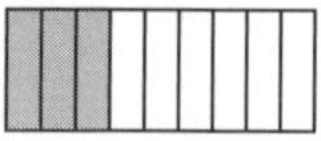

Fractions that are equal to one whole have the same number in the numerator as they do in the denominator. When pictured, the whole pizza, pie, or bar will be shaded.

$$\frac{3}{3} = 1 \quad \frac{5}{5} = 1 \quad \frac{12}{12} = 1 \quad \frac{9}{9} = 1$$

Three Kinds of Fractions

There are three kinds of fractions:

- **Proper fractions:** A proper fraction is one whose numerator is less than the denominator. Examples of proper fractions are $\frac{3}{4}$, $\frac{5}{6}$, $\frac{6}{12}$, and $\frac{5}{15}$.
- **Improper fractions:** An improper fraction is one whose numerator is more than the denominator. Examples of improper fractions are $\frac{6}{4}$, $\frac{12}{3}$, $\frac{7}{4}$, and $\frac{15}{6}$.
- **Mixed number:** A mixed number is a whole number and a proper fraction. Examples of mixed numbers are $4\frac{1}{2}$, $5\frac{4}{6}$, $7\frac{3}{8}$, and $2\frac{2}{16}$.

An improper fraction and a mixed number can be used to name the same amount. You can convert easily between the two fractions.

To change an improper fraction to a mixed number, divide the numerator by the denominator as many times as you can without going over. This becomes the whole number. The remainder becomes the numerator and the denominator from the original fraction stays.

$\frac{14}{3}$ 14 ÷ 3 = 4 with a remainder of 2

$4\frac{2}{3}$

To change from a mixed number to an improper fraction, multiply the denominator by the whole number, add the numerator to the answer you got by multiplying, and write the answer on top of the denominator from the original problem.

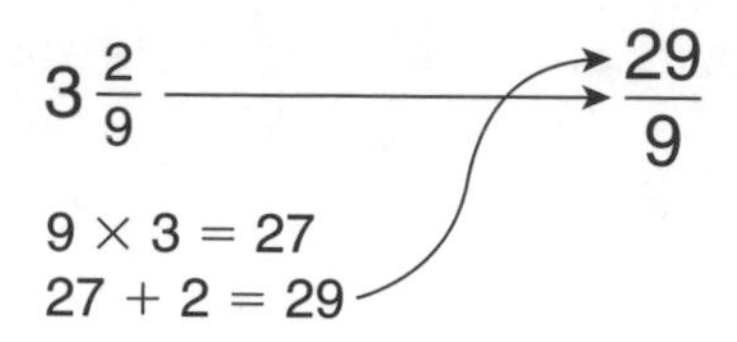

In some cases, when changing from an improper fraction to a mixed number, you might not have a remainder because the number divides in evenly. In this case, there will just be a whole number in the answer. Here are some examples:

$$\frac{14}{7} = 2 \quad \frac{15}{3} = 5 \quad \frac{18}{6} = 3 \quad \frac{24}{3} = 8 \quad \frac{12}{3} = 4$$

An acronym that I use in my classroom to help my students remember the steps to change the mixed number to an improper fraction is MAD:

- **M:** Multiply
- **A:** Add
- **D:** Denominator

A (+)

$4\frac{1}{6}$ ← D – denominator

M (×)

$$4 \times 6 = 24 + 1 = \frac{25}{6}$$

Practice Problems

Convert from an improper fraction to a mixed number.

5.1 $\frac{13}{2} =$

5.2 $\frac{18}{5} =$

5.3 $\frac{23}{8} =$

5.4 $\frac{55}{7} =$

5.5 $\frac{62}{9} =$

Convert from a mixed number to an improper fraction.

5.6 $5\frac{3}{4} =$

5.7 $7\frac{2}{9} =$

5.8 $12\frac{2}{3} =$

5.9 $8\frac{6}{9} =$

5.10 $9\frac{4}{11} =$

Equivalent Fractions

Equivalent fractions are equal in value. They represent the same amount of the whole. You can see in the image that follows that the fractions $\frac{1}{2}$, $\frac{2}{4}$, and $\frac{4}{8}$ are all equivalent. All three fractions are equal to $\frac{1}{2}$ of the pie.

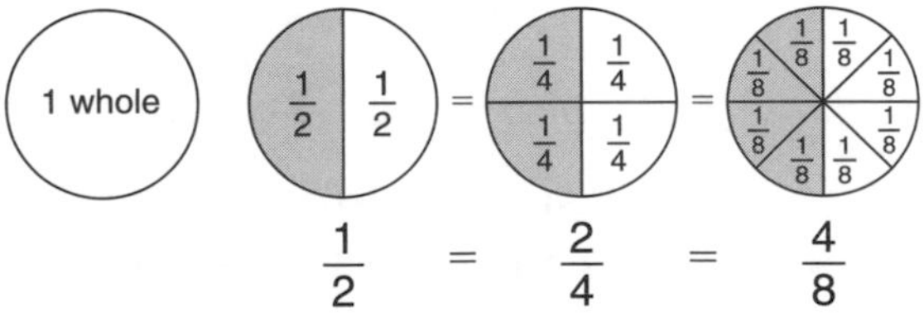

1 whole

$\frac{1}{2} = \frac{2}{4} = \frac{4}{8}$

If you multiply or divide the top and the bottom number by the same number, the result is an equivalent fraction. I always make my students repeat this, "Whatever you do to the BOTTOM, you have to do to the TOP!" For example, $\frac{2}{5}$ is equivalent to $\frac{4}{10}$; both the denominator and the numerator are multiplied by 2.

$$\frac{2 \times 2}{5 \times 2} = \frac{4}{10}$$

Following are some examples of equivalent fractions:

$$\frac{3\,(\times 3)}{8\,(\times 3)} = \frac{9}{24} \qquad \frac{3\,(\times 4)}{8\,(\times 4)} = \frac{12}{32} \qquad \frac{3\,(\times 10)}{8\,(\times 10)} = \frac{30}{80}$$

$$\frac{5\,(\times 2)}{6\,(\times 2)} = \frac{10}{12} \qquad \frac{5\,(\times 3)}{6\,(\times 3)} = \frac{15}{18} \qquad \frac{5\,(\times 5)}{6\,(\times 5)} = \frac{25}{30}$$

You can also divide the top and the bottom to make equivalent fractions. For example, $\frac{12}{20}$ is equivalent to $\frac{6}{10}$; both the numerator and denominator are divided by 2.

$$\frac{12\,(\div 2)}{20\,(\div 2)} = \frac{6}{10}$$

Examples

Here are a few more examples:

$$\frac{24\,(\div 2)}{36\,(\div 2)} = \frac{12}{18} \qquad \frac{24\,(\div 3)}{36\,(\div 3)} = \frac{8}{12} \qquad \frac{24\,(\div 6)}{36\,(\div 6)} = \frac{4}{6}$$

$$\frac{40\,(\div 2)}{50\,(\div 2)} = \frac{20}{25} \qquad \frac{40\,(\div 10)}{50\,(\div 10)} = \frac{4}{5}$$

In some cases, when a problem is given, , only the numerator or only the denominator will be provided. In cases like these, whatever computation has been done to one number needs to be done for the other number.

$$\frac{4\,(\times 5)}{5\,(\times 5)} = \frac{x}{25} \qquad x = 20 \qquad \frac{4}{5} = \frac{20}{25}$$

In this problem, you are given the denominator. You know that $5 \times 5 = 25$ and $4 \times 5 = 20$, so $\frac{4}{5}$ is equivalent to $\frac{20}{25}$.

Following are yet more examples:

$$\frac{4\,(\times 2)}{9\,(\times 2)} = \frac{x}{18} \qquad x = 8 \qquad \frac{4}{9} = \frac{8}{18}$$

$$\frac{5\,(\div 5)}{25\,(\div 5)} = \frac{1}{x} \qquad x = 5 \qquad \frac{5}{25} = \frac{1}{5}$$

Practice Problems

Write 3 equivalent fractions.

5.11 $\frac{2}{6} =$

5.12 $\frac{5}{9} =$

5.13 $\frac{7}{8} =$

5.14 $\frac{6}{9} =$

5.15 $\frac{2}{5} =$

Write the equivalent fraction.

5.16 $\frac{4}{8} = \frac{x}{64}$

5.17 $\frac{2}{6} = \frac{4}{x}$

5.18 $\frac{4}{7} = \frac{x}{21}$

5.19 $\frac{5}{6} = \frac{x}{18}$

5.20 $\frac{1}{8} = \frac{9}{x}$

Simplifying Fractions

Simplifying a fraction means taking it down to the smallest equivalent fraction, also called the lowest term or simplest form. To simplify a fraction, you divide the numerator and the denominator by the greatest common factor (GCF) between the two. At first, it will take children time to simplify fractions. Eventually, as children master math facts and simplify more often, the task will become easier.

A fraction's greatest common factor (GCF) is the greatest factor that will divide into both numbers. To simplify a fraction using the greatest common factor, begin by listing the factors of both numbers, and then look for the largest number that is the same in both lists. This will be used as the GCF to simplify the fraction.

Examples

$\frac{12}{24}$

List the factors of 12 and 24.

12: 1, 2, 3, 4, 6, 12
24: 1, 2, 3, 4, 6, 8, 12, 24

The greatest factor of both 12 and 24 is 12. So, divide both the numerator and the denominator by12. The simplest form or lowest term of $\frac{12}{24}$ is $\frac{1}{2}$.

$$\frac{12\,(\div 12)}{24\,(\div 12)} = \frac{1}{2}$$

$\frac{36}{40}$

List the factors of 36 and 40.

36: 1, 2, 3, 4, 6, 9, 12, 18, 36
40: 1, 2, 4, 5, 8, 10, 20, 40

The greatest common factor of 36 and 40 is 4. Divide both the numerator and denominator by 4. The simplest form of $\frac{36}{40}$ is $\frac{9}{10}$.

$$\frac{36\,(\div 4)}{40\,(\div 4)} = \frac{9}{10}$$

One way to help children list the factors of a number is to make a factor rainbow. You begin by listing the factor and one on the outside of the rainbow as the first factor pair, and then continue to list factor pairs in numerical order. You then connect each factor pair with a curved line to make a rainbow. That way, children can see factors that pair together. This really helps them to not forget any factors of a number.

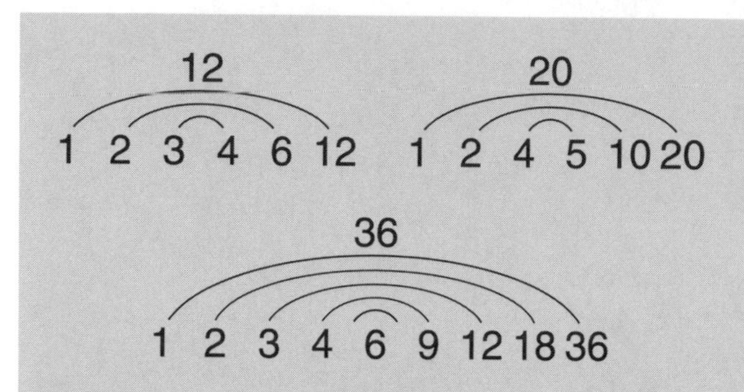

Notice that in the factor rainbow for the number 36, one factor has its own rainbow arch. That is because $6 \times 6 = 36$. The two factors are the same number, so they are only listed one time.

As children simplify fractions more routinely, it will no longer be necessary to list all the factors. They will be able to come up with the GCF just by knowing the factors of numbers.

Practice Problems

Reduce each fraction to simplest form.

5.21 $\frac{6}{9}$

5.22 $\frac{12}{18}$

5.23 $\frac{3}{24}$

5.24 $\frac{5}{25}$

5.25 $\frac{6}{30}$

Comparing Fractions

There are two methods for comparing fractions: cross-multiplying and finding a common denominator. The students in my classroom prefer to use the first method because it is clear and concrete, and done without error most of the time.

Comparing Fractions by Cross-Multiplying

Comparing fractions by cross-multiplying is the easiest way for my students. To compare, you multiply the first fraction's numerator by the second fraction's denominator and then the second fraction's numerator with the first fraction's denominator. After you have multiplied, compare the two products. The larger product tells you which is the larger fraction.

Examples

$\frac{6}{8} \quad \frac{4}{9}$

Begin by writing the two fractions. Next, multiply the first fraction's numerator by the second fraction's denominator and then the second fraction's numerator by the first fraction's denominator.

$6 \times 9 = 54 \qquad 8 \times 4 = 32$

$$\frac{6}{8} \times \frac{4}{9}$$

When multiplying the two numbers on the diagonal, $6 \times 9 = 54$ and $8 \times 4 = 32$.

54 is greater than 32, so $\frac{6}{8}$ is the larger fraction. $\frac{6}{8} > \frac{4}{9}$

$\frac{4}{14} \quad \frac{3}{8}$

Begin by writing the two fractions. Next, multiply the first fraction's numerator by the second fraction's denominator and then the second fraction's numerator by the first fraction's denominator.

$$4 \times 8 = 32 \qquad 14 \times 3 = 42$$

$$\frac{4}{14} \times \frac{3}{8}$$

When multiplying the two numbers on the diagonal, $4 \times 8 = 32$ and $14 \times 3 = 42$.

42 is greater than 32, so $\frac{3}{8}$ is the larger fraction. $\frac{4}{14} < \frac{3}{8}$

Comparing Fractions by Finding a Common Denominator

When two fractions have the same number in the denominator, they are said to have a common denominator. To compare fractions without a common denominator, you must find the least common multiple (LCM). A multiple is a number that you say when you count by that number. Once a common denominator is found it is easy to compare the fractions.

Example

$\frac{2}{7} \quad \frac{1}{3}$

Begin by listing the multiples of each denominator. For this problem, list the multiples of 7 and 3.

7: 7, 14, (21), 28, 35...
3: 3, 6, 9, 12, 15, 18, (21), 24...

The least common number in the two lists is 21. This is the number that will be used to change to common denominators. Write the two fractions with 21 as the common denominator. This is now just like finding equivalent fractions.

$$\frac{2}{7} = \frac{}{21} \qquad \frac{1}{3} = \frac{}{21}$$

The aforementioned saying applies here: "Whatever you do to the BOTTOM, you have to do to the TOP!"

$$\frac{2\,(\times 3)}{7\,(\times 3)} = \frac{6}{21} < \frac{1\,(\times 7)}{3\,(\times 7)} = \frac{7}{21}$$

Once the fractions have been changed to common denominators, the two equivalent fractions are used to compare.

$\frac{6}{21}$ is less than $\frac{7}{21}$, so $\frac{2}{7}$ is less than $\frac{1}{3}$. $\frac{2}{7} < \frac{1}{3}$

Practice Problems

Compare using cross-multiplication or by finding the common denominator.

5.26 $\frac{2}{3}$ $\frac{4}{5}$

5.27 $\frac{6}{8}$ $\frac{2}{9}$

5.28 $\frac{2}{10}$ $\frac{3}{8}$

5.29 $\frac{5}{15}$ $\frac{1}{3}$

5.30 $\frac{8}{9}$ $\frac{11}{12}$

Ordering Fractions

Ordering fractions is a hard concept for children. It is easiest to compare fractions that have like denominators, but that isn't always the case. There are three main approaches for ordering fractions, outlined here.

Ordering Fractions with Like Denominators

Fractions that have like denominators are easy to compare! You simply look at the numerator. The fraction with the largest number in the numerator is the larger fraction.

Example

To order the following fractions from greatest to least, compare the numbers in the numerators.

$$\frac{\mathbf{2}}{9}, \frac{\mathbf{4}}{9}, \frac{\mathbf{7}}{9}, \leftarrow \text{compare the numerators}$$

$$\frac{7}{9}, \frac{4}{9}, \frac{2}{9}$$

The fractions ordered from greatest to least are $\frac{7}{9}, \frac{4}{9}, \frac{2}{9}$.

Ordering Fractions with Like Numerators

Fractions that have the same number in the numerator all have the same number of pieces. In that case, you must look at the denominator to compare the fractions. The fraction with the smallest denominator is the largest fraction. This is because the fraction is broken into a smaller number of total pieces.

Example

To order the following fractions from greatest to least, compare the numbers in the denominator.

$$\frac{2}{\mathbf{3}}, \frac{2}{\mathbf{9}}, \frac{2}{\mathbf{6}}, \leftarrow \text{compare the denominators (the smaller the number, the larger the fraction)}$$

$$\frac{2}{3}, \frac{2}{6}, \frac{2}{9}$$

The fractions ordered from greatest to least are $\frac{2}{3}, \frac{2}{6}, \frac{2}{9}$.

Ordering Fractions with Unlike Denominators

Most times, fractions do not have the same numerators or denominators. In cases like these, you must change the fractions to equivalent fractions with common denominators.

Examples

To order these fractions from greatest to least, change the denominators to common denominators.

$$\frac{3}{4}, \frac{5}{6}, \frac{2}{12} \leftarrow \text{change to common denominator}$$

$$\times 3 \quad \times 2 \quad \times 1$$

$$\frac{9}{12}, \frac{10}{12}, \frac{2}{12} \quad \text{the LCM is 12}$$

Then, compare the numerators to order the fractions.

The fractions ordered from greatest to least are $\frac{5}{6}, \frac{3}{4}, \frac{2}{12}$.

Here's another example. First, change the denominators to common denominators.

$$\frac{2}{8}, \frac{3}{4}, \frac{9}{16} \leftarrow \text{change to common denominator}$$

$$\times 2 \quad \times 4 \quad \times 1$$

$$\frac{4}{16}, \frac{12}{16}, \frac{9}{16} \quad \text{the LCM is 16}$$

Then, compare the numerators to order the fractions.

The fractions ordered from greatest to least are $\frac{3}{4}, \frac{9}{16}, \frac{2}{8}$.

Practice Problems

Order each set of fractions from least to greatest:

5.31 $\frac{2}{6}, \frac{4}{6}, \frac{5}{6}$

5.32 $\frac{5}{6}, \frac{5}{8}, \frac{5}{15}$

5.33 $\frac{2}{7}, \frac{4}{7}, \frac{3}{14}$

5.34 $\frac{1}{8}, \frac{3}{4}, \frac{7}{16}$

5.35 $\frac{2}{3}, \frac{4}{5}, \frac{1}{15}$

Chapter 6

Fraction Computation

Adding and subtracting fractions involves very much the same steps. In both cases, there need to be common denominators, meaning the bottom numbers may need to be changed. I always tell my students this rhyme:

> If adding or subtracting fractions is your game, then
> make sure to keep the denominators the SAME!

This helps them to remember that they need to have common denominators before computing.

Multiplying fractions is the easiest fraction computation because it is done by just multiplying the numerator and the denominator straight across. Common denominators are not needed. Division is done very much the same way as multiplication, but you multiply the reciprocal, or opposite of the second fraction. Once children have the steps down, computation of fractions can be easy to master.

Adding Fractions

When adding fractions with like denominators, you keep the denominators as is and add the numerators only. To add fractions with unlike denominators, you must change one or both of the fractions so they have common denominators. You do this by using the least common multiple (LCM) to change to equivalent fractions.

> If you are unsure how to find the least common denominator among two fractions, see the section "Comparing Fractions by Finding a Common Denominator" in Chapter 5, "Fraction Concepts." This will walk you through finding the LCM. Directions on changing from an improper fraction to a mixed number and reducing a fraction can also be found in Chapter 5. You may need to refer to these rules prior to computing fractions.

The Steps to Add Fractions with Like Denominators

1. Add the numerators of both fractions together. Keep the denominator the same.
2. Reduce or change from an improper fraction if necessary.

Examples

$\frac{3}{6} + \frac{2}{6} = \frac{5}{6}$

Add the numerators together $(3 + 2 = 5)$ and keep the denominator (6). You cannot reduce $\frac{5}{6}$, so that is the final answer.

$\frac{4}{12} + \frac{6}{12} = \frac{10}{12}$ or $\frac{5}{6}$

Add the numerators together $(4 + 6 = 10)$ and keep the denominator (12). You can reduce $\frac{10}{12}$ by dividing both the numerator and denominator by 2, so you can reduce $\frac{10}{12}$ to $\frac{5}{6}$.

$\frac{7}{8} + \frac{4}{8} = \frac{11}{8}$ or $1\frac{3}{8}$

Add the numerators together $(7 + 4 = 11)$ and keep the denominator (8).
$\frac{11}{8}$ is an improper fraction, so it needs to be changed to a mixed number.

$\frac{11}{8}$ becomes $1\frac{3}{8}$.

$$\begin{array}{r} \frac{7}{8} \\ +\frac{4}{8} \\ \hline \frac{11}{8} \end{array} \quad \div \frac{11}{8} \xrightarrow{-8} = 1\frac{3}{8}$$

$\frac{5}{10} + \frac{8}{10} = \frac{13}{10}$ or $1\frac{3}{10}$

$$\begin{array}{r} \frac{5}{10} \\ +\frac{8}{10} \\ \hline \frac{13}{10} \end{array} \quad \div \frac{13}{10} \xrightarrow{-10} = 1\frac{3}{10}$$

Practice Problems

6.1 $\frac{3}{8} + \frac{4}{8} =$

6.2 $\frac{2}{10} + \frac{5}{10} =$

6.3 $\frac{1}{7} + \frac{4}{7} =$

6.4 $\frac{2}{8} + \frac{4}{8} =$

6.5 $\frac{3}{15} + \frac{2}{15} =$

The Steps to Add Fractions with Unlike Denominators

1. Find the least common multiple (LCM) of both denominators.
2. Change the denominators in both fractions to be equivalent fractions with the new denominator.
3. Add the numerators together and keep the denominators the same.
4. Reduce or change from an improper fraction if necessary.

Examples

$$\frac{3}{4}+\frac{1}{8}=\frac{7}{8}$$

The first step is to change the bottom numbers to be common denominators. The LCM of 4 and 8 is 8. You will change the first denominator to 8 by making an equivalent fraction.

$$\begin{array}{lcl} \frac{3\,(\times 2)}{4\,(\times 2)} & = & \frac{6}{8} \\ +\frac{1}{8} & = & \frac{1}{8} \\ \hline \end{array}$$

Add $\frac{6}{8}$ and $\frac{1}{8}$. It is not improper and cannot be reduced, so the final answer is $\frac{7}{8}$.

$$\begin{array}{lcl} \frac{3\,(\times 2)}{4\,(\times 2)} & = & \frac{6}{8} \\ +\frac{1}{8} & = & \frac{1}{8} \\ \hline & & \frac{7}{8} \end{array}$$

add

> One tip that I tell my students is that if the numerator is one less than the denominator, the number cannot be reduced. This is one tip that my students seem to always use and it helps them to quickly decide whether or not a number can be reduced.

Following are more examples.

$\frac{6}{7} + \frac{12}{14} = 1\frac{10}{14}$ or $1\frac{5}{7}$

Change the bottom numbers to be common denominators. The LCM of 7 and 14 is 14. Change the denominator of the first fraction to 14 by making an equivalent fraction.

$$\frac{6\,(\times 2)}{7\,(\times 2)} = \frac{12}{14}$$
$$+\frac{12}{14} = \frac{12}{14}$$

Add $\frac{12}{14}$ and $\frac{12}{14}$. The answer is an improper fraction $\frac{24}{14}$, so it will need to be changed to a proper fraction or mixed number. (See the steps in Chapter 5.)

$$\frac{6\,(\times 2)}{7\,(\times 2)} = \frac{12}{14}$$
$$+\frac{12}{14} = \frac{12}{14}$$
$$\frac{24}{14}$$

add

To change $\frac{24}{14}$ to a mixed number, divide 14 into 24 without going over. The remainder becomes the numerator and the denominator stays the same.

$$\frac{24}{14}$$

$24 \div 14 = 1$ with a remainder of 10

$$1\frac{10}{14}$$

The final answer of $\frac{6}{7} + \frac{12}{14}$ is $1\frac{10}{14}$ or $1\frac{5}{7}$.

$\frac{2}{3} + \frac{1}{5} = \frac{13}{15}$

Change the bottom numbers to be common denominators. The LCM of 3 and 5 is 15. Then change the denominator of both fractions to 15 by making an equivalent fraction.

$$\frac{2\,(\times 5)}{3\,(\times 5)} = \frac{10}{15}$$
$$+\frac{1\,(\times 3)}{5\,(\times 3)} = \frac{3}{15}$$

Add $\frac{10}{15}$ and $\frac{3}{15}$. It is not improper and cannot be reduced, so the final answer is $\frac{13}{15}$.

$$\frac{2\,(\times 5)}{3\,(\times 5)} = \frac{10}{15}$$
$$+\frac{1\,(\times 3)}{5\,(\times 3)} = \frac{3}{15}$$
$$\frac{13}{15}$$

add

Practice Problems

6.6 $\frac{1}{8} + \frac{3}{4} =$

6.7 $\frac{2}{7} + \frac{3}{14} =$

6.8 $\frac{3}{4} + \frac{6}{8} =$

6.9 $\frac{2}{3} + \frac{6}{9} =$

6.10 $\frac{4}{5} + \frac{6}{7} =$

The Steps to Add Mixed Numbers

1. Change the denominators in both fractions to be common denominators.
2. Add the numerators together and keep the denominators the same.
3. Add the whole numbers.
4. Reduce or change from an improper fraction if necessary.

Examples

$2\frac{4}{6} + 1\frac{1}{4} = 3\frac{11}{12}$

The first step is to change the bottom numbers to be common denominators. The LCM of 6 and 4 is 12. You will change both denominators to 12 by making equivalent fractions. Add the numerators and then the whole numbers. $3\frac{11}{12}$ cannot be reduced, so that is the final answer.

$$\begin{array}{rl} 2\frac{4\,(\times 2)}{6\,(\times 2)} = & \mathbf{\frac{8}{12}} \\ +1\frac{1\,(\times 3)}{4\,(\times 3)} = & \mathbf{\frac{3}{12}} \\ \hline 3 & \frac{11}{12} \end{array}$$

$4\frac{3}{7} + 2\frac{9}{14} = 7\frac{1}{14}$

The first step is to change the bottom numbers to be common denominators. The LCM of 7 and 14 is 14. You will change the first denominator to 14 by making an equivalent fraction. Add the numerators and then the whole numbers.

$$\begin{array}{rl} 4\frac{3\,(\times 2)}{7\,(\times 2)} = & \mathbf{\frac{6}{14}} \\ +2\frac{9}{14} = & \mathbf{\frac{9}{14}} \\ \hline 6 & \frac{15}{14} \end{array}$$

$6\frac{15}{14}$ cannot stay as an improper fraction, so it needs to be changed to a mixed number and added to the whole number. $\frac{15}{14}$ is equal to $1\frac{1}{14}$. $1\frac{1}{14} + 6 = 7\frac{1}{14}$.

$$\div \frac{15}{14} \xrightarrow{-14} = 1\frac{1}{14}$$

$$\mathbf{1\frac{1}{14} + 6 = 7\frac{1}{14}}$$

$5\frac{3}{9} + 2\frac{2}{3} = 8$

The first step is to change to common denominators. The LCM of 3 and 9 is 9. Add the numerators and then the whole numbers. $7\frac{9}{9}$ cannot remain an improper fraction. $\frac{9}{9}$ is equal to 1, and $1 + 7 = 8$.

$$\begin{array}{lll} 5\frac{3}{9} & = \frac{3}{9} & \\ +2\frac{2\,(\times 3)}{3\,(\times 3)} & = \frac{6}{9} & \div \frac{9}{9} = 1 + 7 = 8 \\ \hline 7 & \frac{9}{9} & \end{array}$$

$2\frac{3}{4} + 1\frac{5}{6} = 4\frac{7}{12}$

The first step is to change to common denominators. The LCM of 4 and 6 is 12. Add the numerators and then the whole numbers. $3\frac{19}{12}$ cannot remain an improper fraction. $\frac{19}{12}$ is equal to $1\frac{7}{12}$. $1\frac{7}{12} + 3 = 4\frac{7}{12}$.

$$\begin{array}{lll} 2\frac{3\,(\times 3)}{4\,(\times 3)} & = \frac{9}{12} & \\ +1\frac{5\,(\times 2)}{6\,(\times 2)} & = \frac{10}{12} & \div \frac{19}{12} \xrightarrow{-12} = 1\frac{7}{12} + 3 = 4\frac{7}{12} \\ \hline 3 & \frac{19}{12} & \end{array}$$

Practice Problems

6.11 $2\frac{4}{5} + 3\frac{2}{10} =$

6.12 $3\frac{2}{3} + 5\frac{4}{9} =$

6.13 $5\frac{1}{2} + 3\frac{3}{5} =$

6.14 $2\frac{1}{2} + 4\frac{5}{7} =$

6.15 $4\frac{4}{5} + 8\frac{12}{15} =$

Subtracting Fractions

Subtracting follows the same rules as adding fractions. The denominators must match. If they don't, then one or both denominators need to be changed to have common denominators. Remember the rhyme:
"If adding or subtracting fractions is your game, then make sure to keep the denominators the SAME."

The Steps to Subtract Fractions with Like Denominators

1. Subtract the numerators and keep the denominator the same.
2. Reduce or change from an improper fraction if necessary.

Examples

$\frac{7}{12} - \frac{2}{12} = \frac{5}{12}$

Subtract the numerators $(7 - 2 = 5)$ and keep the denominator (12).

$\frac{5}{12}$ cannot be reduced, so that is the final answer.

$\frac{6}{9} - \frac{3}{9} = \frac{3}{9}$

Subtract the numerators (6 – 3 = 3) and keep the denominator (9).

$\frac{3}{9}$ can be reduced by dividing both the numerator and denominator by 3.

$\frac{3}{9}$ can be reduced to $\frac{1}{3}$.

Practice Problems

6.16 $\frac{4}{6} - \frac{1}{6} =$

6.17 $\frac{8}{9} - \frac{4}{9} =$

6.18 $\frac{8}{12} - \frac{2}{12} =$

6.19 $\frac{15}{16} - \frac{10}{16} =$

6.20 $\frac{11}{20} - \frac{7}{20} =$

The Steps to Subtract Fractions with Unlike Denominators

1. Find the least common multiple (LCM) of both denominators.
2. Change the denominators in both fractions to be equivalent fractions with the new denominator.
3. Subtract the numerators and keep the denominators the same.
4. Reduce if necessary.

Examples

$\frac{3}{4} - \frac{2}{8} = \frac{4}{8}$ or $\frac{1}{2}$

The first step is to change the bottom numbers to be common denominators. The LCM of 4 and 8 is 8. You change the first denominator to 8 by making an equivalent fraction. $\frac{6}{8} - \frac{2}{8} = \frac{4}{8}$. You can reduce $\frac{4}{8}$ by dividing both the numerator and denominator by 4. $\frac{4}{8}$ can be reduced to $\frac{1}{2}$.

$$\frac{3\,(\times 2)}{4\,(\times 2)} = \frac{6}{8}$$
$$-\frac{2}{8} = \frac{2}{8}$$
$$\frac{4\,(\div 4)}{8\,(\div 4)} = \frac{1}{2}$$

$\frac{8}{9} - \frac{2}{6} = \frac{10}{18}$ or $\frac{5}{9}$

The first step is to change the bottom numbers to be common denominators. The LCM of 9 and 6 is 18. You change both denominators to 18 by making equivalent fractions. $\frac{16}{18} - \frac{6}{18} = \frac{10}{18}$. You can reduce $\frac{10}{18}$ by dividing both the numerator and denominator by 2. $\frac{10}{18}$ can be reduced to $\frac{5}{9}$.

$$\frac{8\,(\times 2)}{9\,(\times 2)} = \frac{16}{18}$$
$$-\frac{2\,(\times 3)}{6\,(\times 3)} = \frac{6}{18}$$
$$\frac{10\,(\div 2)}{18\,(\div 2)} = \frac{5}{9}$$

$\frac{9}{15} - \frac{2}{5} = \frac{3}{15}$ or $\frac{1}{5}$

$$\begin{array}{lcl} \frac{9}{15} & = & \frac{9}{15} \\ -\frac{2\ (\times 3)}{5\ (\times 3)} & = & \frac{6}{15} \\ \hline & & \frac{3\ (\div 3)}{15\ (\div 3)} = \frac{1}{5} \end{array}$$

Practice Problems

6.21 $\frac{3}{4} - \frac{7}{12} =$

6.22 $\frac{1}{2} - \frac{2}{8} =$

6.23 $\frac{11}{12} - \frac{5}{6} =$

6.24 $\frac{8}{10} - \frac{2}{5} =$

6.25 $\frac{4}{5} - \frac{2}{3} =$

The Steps to Subtract Mixed Numbers

1. Change the denominators in both fractions to be common denominators.
2. Subtract the numerators. If the fraction on top is smaller than the fraction on the bottom, borrow a 1 from the whole number and change it into a fraction.
3. Subtract the whole numbers.
4. Reduce or change from an improper fraction if necessary.

Examples

$4\frac{8}{10} - 2\frac{3}{5} = 2\frac{2}{10}$ or $2\frac{1}{5}$

The first step is to change the bottom numbers to be common denominators. The LCM of 10 and 5 is 10. You will change the second denominator to 10 by making an equivalent fraction. Subtract the numerators and then the whole numbers. You can reduce $2\frac{2}{10}$ by dividing the numerator and denominator by 2. $\frac{2}{10}$ can be reduced to $\frac{1}{5}$.

$$\begin{array}{lcl} 4\frac{8}{10} & = & \frac{8}{10} \\ -2\frac{3\,(\times 2)}{5\,(\times 2)} & = & \frac{6}{10} \\ \hline 2 & & \frac{2\,(\div 2)}{10\,(\div 2)} = \frac{1}{5} \end{array}$$

$6\frac{4}{5} - 2\frac{3}{7} = 4\frac{13}{35}$

The first step is to change the bottom numbers to be common denominators. The LCM of 5 and 7 is 35. You will change both denominators to 35 by making equivalent fractions. Subtract the numerators and then the whole numbers. $\frac{13}{35}$ cannot be reduced, so the final answer is $4\frac{13}{35}$.

$$\begin{array}{lcl} 6\frac{4\,(\times 7)}{5\,(\times 7)} & = & \frac{28}{35} \\ -2\frac{3\,(\times 5)}{7\,(\times 5)} & = & \frac{15}{35} \\ \hline 4 & & \frac{13}{35} \end{array}$$

$7\frac{1}{8} - 3\frac{4}{16} = 3\frac{14}{16}$ or $3\frac{7}{8}$

The first step in this problem is to change the bottom numbers to be common denominators. The LCM of 8 and 16 is 16. You will change the first denominator to 16 by making an equivalent fraction. In this problem you cannot subtract $\frac{2}{16} - \frac{4}{16}$ because the fraction on the top is smaller.

$$\begin{array}{r} 7\frac{1\,(\times 2)}{8\,(\times 2)} = \frac{2}{16} \\ -3\frac{4}{16} \quad = \frac{4}{16} \\ \hline \end{array}$$

STOP! You cannot subtract 4 from 2!

The next step is to take one whole from the whole number and add it to $\frac{2}{16}$. When regrouping fractions, take one whole and convert that whole to a fraction using the common denominator. In this case, one whole is equal to $\frac{16}{16}$. Add the two fractions to get a new fraction on top: $\frac{2}{16} + \frac{16}{16} = \frac{18}{16}$.

$$\begin{array}{r} 6\not{7}\frac{1\,(\times 2)}{8\,(\times 2)} = \frac{2}{16} + \frac{16}{16} = \frac{18}{16} \\ -3\frac{4}{16} \quad = \frac{4}{16} \qquad\qquad\quad \\ \hline \end{array}$$

Once you have a larger fraction on top, subtract the fractions and then the whole numbers: $\frac{18}{16} - \frac{4}{16} = \frac{14}{16}$ and $6 - 3 = 3$. The answer is $3\frac{14}{16}$, which you can reduce by dividing the 14 and 16 by 2. The final answer is $3\frac{7}{8}$.

$$\begin{array}{r} 6\not{7}\frac{1\,(\times 2)}{8\,(\times 2)} = \frac{2}{16} + \frac{16}{16} = \frac{18}{16} \\ -3\frac{4}{16} \quad = \frac{4}{16} \qquad\qquad\quad \\ \hline 3 \qquad \frac{14\,(\div 2)}{16\,(\div 2)} = 3\frac{7}{8} \qquad\quad \end{array}$$

$$12\frac{1}{5} - 9\frac{3}{4} = 2\frac{9}{20}$$

The first step in this problem is to change the bottom numbers to be common denominators. The LCM of 4 and 5 is 20. You will change both denominators to 20 by making equivalent fractions. In this problem, you cannot subtract $\frac{4}{20} - \frac{15}{20}$ because the fraction on the top is smaller.

$$\begin{array}{r} 12\frac{1\,(\times4)}{5\,(\times4)} = \frac{4}{20} \\ -9\frac{3\,(\times5)}{4\,(\times5)} = \frac{15}{20} \\ \hline \end{array}$$

STOP!
You cannot
subtract 15 from 4!

The next step is to take one whole from the whole number and add it to $\frac{4}{20}$. When regrouping fractions, take one whole and convert that whole to a fraction using the common denominator. In this case, one whole is equal to $\frac{20}{20}$. Add the two fractions to get a new fraction on top: $\frac{4}{20} + \frac{20}{20} = \frac{24}{20}$.

$$\begin{array}{r} 11\not{12}\frac{1\,(\times4)}{5\,(\times4)} = \frac{4}{20} + \frac{20}{20} = \frac{24}{20} \\ -9\frac{3\,(\times5)}{4\,(\times5)} = \frac{15}{20} \\ \hline \end{array}$$

Once you have a larger fraction on top, subtract the fractions and then the whole numbers: $\frac{24}{20} - \frac{15}{20} = \frac{9}{20}$ and $11 - 9 = 2$. The answer is $2\frac{9}{20}$, which cannot be reduced, so that is the final answer.

$$\begin{array}{r} 11\not{12}\frac{1\,(\times4)}{5\,(\times4)} = \frac{4}{20} + \frac{20}{20} = \frac{24}{20} \\ -9\frac{3\,(\times5)}{4\,(\times5)} = \frac{15}{20} \\ \hline 2\frac{9}{20} \end{array}$$

Practice Problems

6.26 $5\frac{3}{4} - 2\frac{1}{3} =$

6.27 $4\frac{4}{6} - 2\frac{1}{12} =$

6.28 $7\frac{5}{6} - 4\frac{2}{5} =$

6.29 $6\frac{1}{5} - 2\frac{2}{3} =$

6.30 $8\frac{2}{4} - 3\frac{8}{10} =$

Multiplying Fractions

Multiplying simple fractions is the easiest of the four types of computations with fractions because you simply multiply straight across. There is no finding common denominators or regrouping when you do the computation. When multiplying mixed numbers, the mixed number is changed to an improper fraction and then multiplied straight across. Just three simple steps and you are done!

The Steps to Multiply Fractions

1. Multiply the numerators.
2. Multiply the denominators.
3. Reduce the fraction if needed.

Reducing fractions is the hardest part of multiplying fractions. It is much easier to reduce on the diagonal before you do the computation. This saves you the step of reducing at the end. I have my students use a highlighter to highlight the numbers on the diagonals. You can also circle them, as shown here. Reduce fractions first like this:

1. Find a number that will divide into both numbers on the diagonal.
2. Divide both numbers by that number to reduce the fraction.
3. Multiply the numerator and denominator. The answer should already be reduced to the lowest terms, but if not, reduce lower if necessary.

$$\frac{\overset{1}{\cancel{4}}}{\underset{1}{\cancel{5}}} \times \frac{\overset{2}{\cancel{10}}}{\underset{3}{\cancel{12}}} = \frac{1}{1} \times \frac{2}{3} = \frac{2}{3} \quad (\div 5,\ \div 4)$$

Examples

$\frac{4}{5} \times \frac{2}{3} = \frac{8}{15}$

Multiply the numerators ($4 \times 2 = 8$) and the denominators ($5 \times 3 = 15$). The final answer is $\frac{8}{15}$, which cannot be reduced.

$\frac{6}{7} \times \frac{2}{4} = \frac{12}{28}$ or $\frac{3}{7}$

Multiply the numerators ($6 \times 2 = 12$) and the denominators ($7 \times 4 = 28$). The result, $\frac{12}{28}$, can be reduced. To do so, divide the numerator and denominator by 4. $\frac{12}{28}$ can be reduced to $\frac{3}{7}$.

$\frac{4}{9} \times \frac{3}{7} = \frac{12}{63}$ or $\frac{4}{21}$

Multiply the numerators ($4 \times 3 = 12$) and the denominators ($9 \times 7 = 63$). You can reduce the result, $\frac{12}{63}$, by dividing the numerator and denominator by 3. $\frac{12}{63}$ can be reduced to $\frac{4}{21}$. The following problem shows where reducing the fraction on the diagonal prior to multiplying is easier than waiting until the end.

$$\div 3 \quad \frac{4}{\cancel{9}_{3}} \times \frac{\cancel{3}^{1}}{7} = \frac{4}{3} \times \frac{1}{7} = \mathbf{\frac{4}{21}}$$

Practice Problems

6.31 $\frac{2}{8} \times \frac{4}{5} =$

6.32 $\frac{4}{6} \times \frac{3}{9} =$

6.33 $\frac{4}{16} \times \frac{2}{5} =$

6.34 $\frac{3}{8} \times \frac{16}{24} =$

6.35 $\frac{7}{12} \times \frac{6}{8} =$

The Steps to Multiply Mixed Numbers

1. Change the mixed number or whole number to an improper fraction.
2. Multiply the numerators and denominators.
3. Convert an answer that may be an improper fraction back to a mixed number.
4. Reduce the fraction if needed. (Don't forget you can reduce on the diagonal to make it easier to reduce the fraction. See the preceding parent tip for steps.)

Examples

$\frac{2}{6} \times 5 = \frac{10}{6}$ or $1\frac{2}{3}$

In this problem, you are multiplying a fraction by a whole number. You need to change the whole number to an improper fraction by placing it over 1. So 5 becomes $\frac{5}{1}$. Then, multiply $\frac{2}{6} \times \frac{5}{1} = \frac{10}{6}$.

$\frac{2}{6} \times \frac{5}{1} = \mathbf{\frac{10}{6}}$

$\frac{10}{6}$ is an improper fraction and needs to be changed to a mixed number.

$\frac{10}{6}$ is $1\frac{4}{6}$ and can be reduced by dividing the numerator and denominator by 2. The final answer is $1\frac{2}{3}$.

$$\frac{10}{6} \overset{-6}{=} 1\frac{4\,(\div 2)}{6\,(\div 2)} = \mathbf{1\frac{2}{3}}$$

(change to a mixed number) (reduce)

$1\frac{2}{3} \times 2\frac{4}{5} = 4\frac{2}{3}$

The first step is to change the mixed numbers to improper fractions.

$1\frac{2}{3}$ becomes $\frac{5}{3}$ and $2\frac{4}{5}$ changes to $\frac{14}{5}$.

$$1\frac{2}{3} \times 2\frac{4}{5} =$$

$$\frac{5}{3} \times \frac{14}{5} =$$

Next, reduce the fractions on the diagonals by dividing both the numerator and denominator by 5. Then multiply the numerators and the denominators.

$$\frac{\overset{1}{\cancel{5}}}{3} \times \frac{14}{\underset{1}{\cancel{5}}} \quad \div 5$$

$$\frac{1}{3} \times \frac{14}{1} = \mathbf{\frac{14}{3}}$$

$\frac{14}{3}$ is an improper fraction and needs to be changed to a mixed number.

The final answer is $4\frac{2}{3}$.

$$\frac{1}{3} \times \frac{14}{1} = \mathbf{\frac{14}{3}} \quad \frac{14}{3} \overset{-12}{=} \mathbf{4\frac{2}{3}}$$

$2\frac{2}{8} \times 3\frac{1}{5} = 7\frac{1}{5}$

The first step is to change the mixed numbers to improper fractions. $2\frac{2}{8}$ becomes $\frac{18}{8}$ and $3\frac{1}{5}$ changes to $\frac{16}{5}$.

$$2\frac{2}{8} \times 3\frac{1}{5} =$$

$$\frac{18}{8} \times \frac{16}{5} =$$

Next, reduce the fractions on the diagonals by dividing both the numerator and denominator by 8. Then multiply the numerators and the denominators.

$$\div 8 \quad \frac{18}{\cancel{8}_{1}} \times \frac{\cancel{16}^{2}}{5} =$$

$$\frac{18}{1} \times \frac{2}{5} = \mathbf{\frac{36}{5}}$$

$\frac{36}{5}$ is an improper fraction and needs to be changed to a mixed number.

The final answer is $7\frac{1}{5}$.

$$\frac{18}{1} \times \frac{2}{5} = \mathbf{\frac{36}{5}} \quad \div \frac{36}{5} \overset{-35}{=} \mathbf{7\frac{1}{5}}$$

Practice Problems

6.36 $3\frac{2}{5} \times 2\frac{2}{4} =$

6.37 $2\frac{3}{7} \times 1\frac{2}{5} =$

6.38 $4\frac{2}{3} \times 1\frac{2}{7} =$

6.39 $6\frac{3}{10} \times 1\frac{1}{9} =$

6.40 $3\frac{1}{9} \times 2\frac{4}{7} =$

Dividing Fractions

Dividing fractions follows the same steps as multiplying fractions. The only difference is that you multiply the reciprocal of the second fraction. The reciprocal of a fraction is the fraction flipped upside down.

The Steps to Divide Fractions

1. Change the second fraction (the one you are dividing by) to the reciprocal. For example, $\frac{3}{4}$ would become $\frac{4}{3}$.
2. Multiply the first fraction by the reciprocal.
3. Reduce the fraction or change from an improper fraction to a mixed number if necessary.

$\frac{4}{5}$	$\div$	$\frac{1}{2}$
leave	**change**	**reciprocal**
↓	↓	↓
$\frac{4}{5}$	$\times$	$\frac{2}{1}$

Examples

$\frac{4}{7} \div \frac{4}{5} = \frac{5}{7}$

The first step is to change the second fraction to the reciprocal.

$\frac{4}{5}$ becomes $\frac{5}{4}$. Then change the problem to a multiplication problem and multiply the reciprocal.

$\frac{4}{7} \div \frac{4}{5}$

$\frac{4}{7} \times \frac{\mathbf{5}}{\mathbf{4}}$

Next, reduce the fractions on the diagonals by dividing both the numerator and denominator by 4. Then multiply the numerators and denominators. The final answer is $\frac{5}{7}$.

$\frac{\cancel{4}^{1}}{7} \times \frac{5}{\cancel{4}_{1}} \div 4 \quad \frac{1}{7} \times \frac{5}{1} = \frac{\mathbf{5}}{\mathbf{7}}$

$\frac{4}{5} \div \frac{2}{5} = 2$

The first step is to change the second fraction to the reciprocal. $\frac{2}{5}$ becomes $\frac{5}{2}$. Then change the problem to a multiplication problem and multiply the reciprocal.

$\frac{4}{5} \div \frac{2}{5}$

$\frac{4}{5} \times \frac{\mathbf{5}}{\mathbf{2}}$

Next, reduce the fractions on the diagonals. In this problem, you can reduce on both diagonals by dividing by 5 and 2. The final answer is 2.

$\frac{\cancel{4}^{2}}{\cancel{5}_{1}} \times \frac{\cancel{5}^{1}}{\cancel{2}_{1}} \begin{matrix}\div 5 \\ \div 2\end{matrix} \quad \frac{2}{1} \times \frac{1}{1} = \frac{2}{1} = \mathbf{2}$

$$\frac{5}{8} \div 2 = \frac{5}{16}$$

When dividing a fraction by a whole number, change the whole number into a fraction ($2 = \frac{2}{1}$). Then change the problem to a multiplication problem and multiply the reciprocal of that fraction.

$$\frac{5}{8} \div 2 = \frac{5}{8} \div \frac{2}{1}$$

$$\frac{5}{8} \times \frac{1}{2}$$

The fractions cannot be reduced on the diagonal, so just multiply the numerators and the denominators. The final answer is $\frac{5}{16}$.

$$\frac{5}{8} \times \frac{1}{2} = \frac{5}{16}$$

Practice Problems

6.41 $\frac{3}{4} \div \frac{1}{2} =$

6.42 $\frac{1}{2} \div \frac{2}{9} =$

6.43 $\frac{2}{3} \div 7 =$

6.44 $\frac{5}{9} \div 4 =$

6.45 $\frac{2}{3} \div \frac{4}{5} =$

The Steps to Divide Mixed Numbers

1. Change each mixed number to an improper fraction.
2. Change the second fraction (the one you are dividing by) to the reciprocal. For example, $\frac{3}{4}$ would become $\frac{4}{3}$.
3. Multiply the first fraction by the reciprocal.
4. Reduce the fraction or change from an improper fraction to a mixed number if necessary.

Examples

$4\frac{2}{4} \div \frac{3}{4} = \frac{18}{3}$ or 6

The first step is to change $4\frac{2}{4}$ to an improper fraction ($\frac{18}{4}$). Next, change the second fraction to the reciprocal. Then change the problem to a multiplication problem and multiply the reciprocal.

$$4\frac{2}{4} \div \frac{3}{4}$$

$$\frac{18}{4} \times \frac{\mathbf{4}}{\mathbf{3}}$$

Next, reduce the fractions on the diagonals by dividing both the numerator and denominator by 4. Then multiply the numerators and denominators. Change the improper fraction $\frac{18}{3}$ to a mixed number, or in this case a whole number. The final answer is 6.

$$\frac{18}{\not{4}_{1}} \times \frac{\not{4}^{1}}{3} \div 4 \qquad \frac{18}{1} \times \frac{1}{3} = \frac{\mathbf{18}}{\mathbf{3}}$$

$$\div \frac{18}{3} = \mathbf{6}$$

$2\frac{1}{2} \div 3\frac{1}{8} = \frac{4}{5}$

The first step is to change $2\frac{1}{2}$ and $3\frac{1}{8}$ to improper fractions ($\frac{5}{2}$ and $\frac{25}{8}$).

Next, change the second fraction to the reciprocal. $\frac{25}{8}$ becomes $\frac{8}{25}$.

Then change the problem to a multiplication problem and multiply the reciprocal.

$$2\frac{1}{2} \div 3\frac{1}{8}$$

$$\frac{5}{2} \div \frac{25}{8}$$

$$\frac{5}{2} \times \frac{\mathbf{8}}{\mathbf{25}}$$

Next, reduce the fractions on the diagonals. In this problem, you can reduce on both diagonals by dividing by 5 and 2. Then multiply the numerators and denominators. The final answer is $\frac{4}{5}$.

$$\frac{\overset{1}{\cancel{5}}}{\underset{1}{\cancel{2}}} \times \frac{\overset{4}{\cancel{8}}}{\underset{5}{\cancel{25}}} \quad \div 2,\ \div 5 \qquad \frac{1}{1} \times \frac{4}{5} = \frac{\mathbf{4}}{\mathbf{5}}$$

Practice Problems

6.46 $3\frac{1}{9} \div \frac{4}{5} =$

6.47 $2\frac{1}{4} \div \frac{3}{4} =$

6.48 $2\frac{2}{3} \div 1\frac{1}{3} =$

6.49 $3\frac{1}{2} \div 2\frac{1}{3} =$

6.50 $3\frac{5}{8} \div 1\frac{1}{8} =$

Chapter 7

Decimals, Fractions, and Percents

This chapter will walk you through the conversion among decimals, fractions, and percents, as well as through many decimal concepts.

Decimals

A *decimal* is basically another way to write a fraction. A decimal is part of a whole. The decimal point is placed to the right of the ones place in a number. The decimal point is the most important part of the decimal because it shows you exactly what place value each number is worth in the number. Any number to the left of the decimal point represents a number greater than one, or a whole number. Any number to the right of the decimal point represents a number smaller than one. One concept that is crucial to success with decimals is knowing the value of the places. If this is an issue, use the following chart to work with your child on this concept.

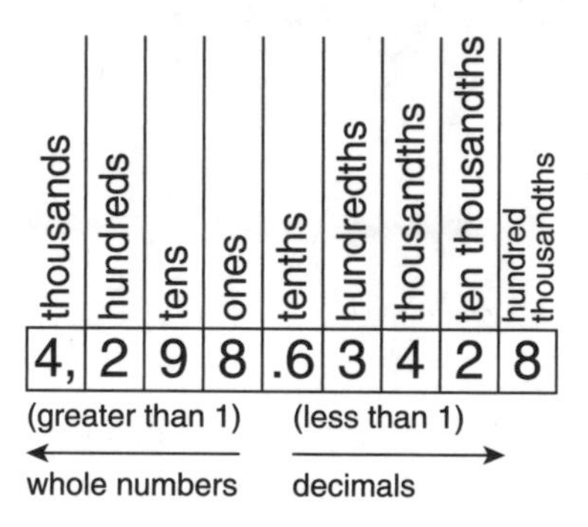

Comparing Decimals

Comparing decimals is one concept that you will use throughout your life, especially when dealing with money. It is very easy to compare decimals just by looking at the whole numbers to the left of the decimal point. If they are the same, then it is done by comparing the decimals on the right. It is important to use the decimal point when comparing decimals because this tells you the value of the digits that you will be comparing.

The Steps to Compare Decimals

1. Compare the numbers to the left of the decimal point, the whole numbers. If they are different, the bigger number has a greater value.

2. If the whole numbers are the same, compare the numbers to the right of the decimal. The bigger number is the decimal that has a greater value. If the numbers to the right of the decimal point do not have the same number of places, add zeros to make the number of places the same. This does not change the value of the number.

Examples

214.5_____241.6

In this problem, the numbers to the left of the decimal point, or the whole numbers, are different. Circle the whole numbers and then compare them. 214 is less than 241, so $214.5 < 241.6$.

214.5 __ 241.6

$214 < 241$
(less than)

$214.5 \underline{<} 241.6$

256.52_____256.413

In this problem, the numbers to the left of the decimal point, or the whole numbers, are the same. The next step is to circle and compare the numbers to the right of the decimal point. Notice that there is not the same number of decimal places in each number, making them difficult to compare.

256.52 __ 256.413

Add a zero to the 52 to make it 520. Adding a zero does not change the value of the decimal. You can now compare the two numbers because they have the same number of decimal places. Circle the decimals and then compare them. 520 is greater than 413, so $256.52 > 256.413$.

$256.520 \underline{>} 256.413$

5.874_____5.81

As with the preceding problem, the numbers to the left of the decimal point, or the whole numbers, are the same. The next step is to circle and compare the numbers to the right of the decimal point. Notice that there is not the same number of decimal places in each number, making them difficult to compare.

5.874 __ 5.81

Add a zero to the 81 to make it 810. Adding a zero does not change the value of the decimal. You can now compare the two numbers because they have the same number of decimal places. Circle the decimals and then compare them. 874 is greater than 810, so 5.874 > 5.81.

5.874 > 5.810

If your child is having a hard time figuring out how to write the greater-than sign (>) and the less-than sign (<), teach them this saying: The alligator's mouth eats the bigger number. Students in my classroom who struggle like to draw the teeth in the sign to help them remember.

1.9 > 0.6

Practice Problems

Indicate which decimal is greater.

7.1 56.21______65.21

7.2 420.3_______42.358

7.3 25.85_______25.853

7.4 142.9_______142.96

7.5 42.987_______42.99

Ordering Decimals

Ordering decimals is an easy concept to grasp if your child has a solid foundation of the place values of whole numbers and decimals. If your child is struggling to grasp place value, use the chart at the beginning of this chapter while ordering decimals. Knowing that ascending means "going up," or ordering from least to greatest, and descending means "going down," or ordering from greatest to least, will help with directions.

The Steps to Order Decimals

1. Write the numbers in a column, one below the other, lining up the decimal points.
2. Add zeros to the right of the last digit if necessary to make all numbers have the same number of decimal places.
3. Compare the decimals from the largest place value digit. If you are comparing the decimals in ascending order (least to greatest), you will order the decimals by looking for the smallest number in each place value. If comparing the decimals in descending order (greatest to least), you will compare by finding the largest number in each place value.

Examples

Order these decimals from least to greatest:

2.356, 2.41, 0.74, 1.85

First, write the decimals in a column, lining up the decimal points.

2.356
2.41
0.74
1.85

Next, add a zero(s) to the right of each number as needed so they have the same number of decimal places.

2.356
2.41**0**
0.74**0**
1.85**0**

Compare the decimals from least to greatest. Begin by comparing the number in the greatest place value—in this case, the ones place. 0.740 is the smallest decimal because it has a 0 in the ones place. Draw a little 1 to the side of the decimal. 1.850 is the next in order because it has a 1 in the ones place. Draw a little 2 by it. The last two decimals each have a 2 in the ones place, so move to the next greatest place value, the tenths, to compare. 2.356 has a 3 in the tenths place, so it is the next in order, followed by 2.410. Draw a little 3 by 2.356 and a little 4 by 2.410.

$2.\underline{3}56$ ③
$2.\underline{4}10$ ④
0.740 ①
1.850 ②

The decimals ordered from least to greatest are as follows:

1. 0.74
2. 1.85
3. 2.356
4. 2.41

Order these decimals from greatest to least:

5.41, 5.842, 5.9, 4.741

First, write the decimals in a column, lining up the decimal points.

5.41
5.842
5.9
4.741

Next, add a zero(s) to the right of each number as needed so they have the same number of decimal places.

5.410
5.842
5.900
4.741

Then compare the decimals from greatest to least. Begin by comparing the number in the greatest place value—in this case, the ones place. There are three decimals that have a 5 in the ones place, so move to the next greatest place value, the tenths, to compare. 5.900 has a 9 in the tenths, so it is the greatest decimal, followed by 5.842 and 5.410. The smallest decimal is 4.741.

5.$\underline{4}$10 ③
5.$\underline{8}$42 ②
5.$\underline{9}$00 ①
4.741 ④

The decimals ordered from greatest to least are as follows:

1. 5.9
2. 5.842
3. 5.41
4. 4.741

Practice Problems

Order from greatest to least:

7.6 0.5, 4.2, 4.58, 3.41

7.7 7.44, 6.31, 6.13, 6.014

7.8 12.5, 12.81, 12.99, 12.7

Order from least to greatest:

7.9 1.8, 0.52, 0.41, 0.8

7.10 3.1, 3.54, 3.852, 3.74

Rounding Decimals

Rounding decimals is a way of estimating a number. This is a concept that will be used frequently in the real world, whether it be with money, time, or distance traveled.

One visual that I use with my students to help them decide whether to round is a car stuck on top of a hill, as shown here. The engine is heavier than the trunk, so at 5, the car will fall forward. Similarly, if a decimal is greater than 5, it will be rounded up. Anything less than 5 will cause the car to fall backward. Similarly, if a decimal is less than 5, it will round down.

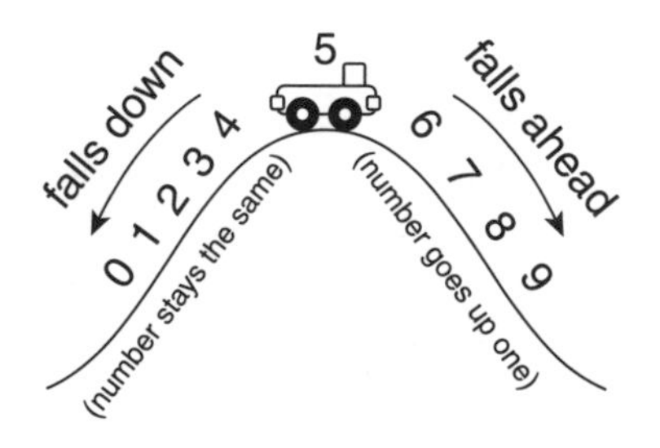

Using this visual, if the decimal you are looking at is a 0, 1, 2, 3, or 4, the number in the place to its left will stay the same. If the decimal is a 5, 6, 7, 8, or 9, the number in the place to its left will go up one. This helps children to remember which numbers stay the same and which will go up one. A saying that also helps my students remember when to change the number or round up is, “Five or more, raise the score. Four or less, give it a rest.” This will give your child another mental reminder of when to round up or keep the number the same.

The Steps to Round Decimals

1. Underline the place in the number that is being rounded.
2. Put an arrow under the digit to the right of the underlined number. This is the number you will use to help you round.
3. If the number to which the arrow is pointing is a 0, 1, 2, 3, or 4, the number you underlined will stay the same, and all the digits to the right of the number will be dropped. If the number to which the arrow is pointing is a 5, 6, 7, 8, or 9, you add one to the underlined number and drop all the digits to the right of the number.

Examples

Round 56.76 to the nearest tenth.

First, underline the digit in the tenths place, the 7, and place an arrow under the digit to the right, the 6. The 6 tells you whether to keep the 7 the same or to add one.

56.76

The digit to the right of the tenths place is a 6, so add one to the 7 and drop all the digits to the right of the tenths place (the place being rounded).

56.76 = 56.8

Round 25.482 to the nearest hundredth.

First, underline the digit in the hundredths place, the 8, and place an arrow under the digit to the right, the 2. The 2 tells you whether to keep the 8 the same or to add one.

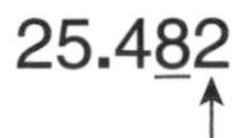

The digit to the right of the tenths place is a 2, so the 8 will stay the same and drop all the digits to the right of the hundredths place (the place being rounded).

$$25.4\underline{8}2 = 25.48$$

Round 14.852 to the ones place. (Sometimes the directions for rounding to the ones place will say, "Round to the nearest whole number.")

First, underline the digit in the ones place, the 4, and place an arrow under the digit to the right, the 8. The 8 tells you whether to keep the 4 the same or to add one.

$$1\underline{4}.852$$

The digit to the right of the ones place is an 8, so add one to the 4 and drop all the digits to the right of the ones. In this case, there will not be any decimal numbers.

$$1\underline{4}.852 = 15$$

Practice Problems

Round to the nearest tenth.

7.11 23.45

7.12 148.963

7.13 74.258

Round to the nearest hundredth.

7.14 52.852

7.15 4.738

Converting Decimals, Fractions, and Percents

Many decimals are easy to memorize. In such cases, children can simply memorize how to convert among fractions, decimals, and percents. But for those that are not easy to memorize, there are some easy conversion rules that will help you convert. All three—decimals, fractions, and percents—describe part of a whole. Once your child grasps that concept, it is easy to master.

The Steps to Convert from a Decimal to a Fraction

1. Read the decimal and decide if the decimal goes to the tenths place, the hundredths place, the thousandths place, etc.
2. Write the decimal as a fraction over 10, 100, 1,000, and so on.
3. If necessary, reduce the fraction to the lowest terms.

Examples

0.8

When spoken aloud, "0.8" is said as "eight tenths." So you know the decimal goes to the tenths place. Place 8 over 10 to become $\frac{8}{10}$. Then divide the numerator and the denominator by 2 to find the fraction in its simplest form: $\frac{4}{5}$.

0.8 eight tenths $\frac{8}{10}$ or $\frac{4}{5}$

0.42

When spoken aloud, "0.42" is said as "forty-two hundredths." So you know the decimal goes to the hundredths place. Place 42 over 100 to become $\frac{42}{100}$. Divide the numerator and the denominator by 2 to find the fraction in its simplest form: $\frac{21}{50}$.

0.42 forty-two hundredths $\frac{42}{100}$ or $\frac{21}{50}$

1.7

When spoken aloud, "1.7" is said "one and seven tenths." So you know there is one whole number and that the 7 is in the tenths place. Place 7 over 10 to become $\frac{7}{10}$. Then place the 1 in front, to make $1\frac{7}{10}$.

1.7 one and seven tenths $1\frac{7}{10}$

Practice Problems

Convert from a decimal to a fraction:

7.16 0.6

7.17 0.54

7.18 0.80

7.19 1.23

7.20 2.6

The Steps to Convert from a Decimal to a Percent

1. Write the number as a decimal.
2. Move the decimal two places to the right. You are multiplying the decimal by 100.
3. Add the percent sign (%) to show that the number is a percent.

Examples

0.56 = 56%

Move the decimal two places to the right and add a percent sign (%).

0.56 = 56%

0.6 = 60%

Move the decimal two places to the right. Place a zero in the open place and add a percent sign (%).

0.6_ = 60%

add a zero to
the open space

3.25 = 325%

Move the decimal two places to the right and add a percent sign (%). Notice in this problem that there is a whole number in the decimal. This means you have more than one whole, thus the percent will be larger than 100.

3.25 = 325%

Practice Problems

Convert from a decimal to a percent:

7.21 0.4

7.22 0.29

7.23 0.46

7.24 2.54

7.25 1.6

The Steps to Convert from a Percent to a Decimal

1. Write the number as a percent. The decimal is hiding behind the number in the ones place (behind the number farthest to the right).

2. Move the decimal two places to the left. You are dividing the percent by 100 because *percent* means *per 100*.

Examples

76% = 0.76

Move the decimal two places to the left.

76%   76.% = 0.76

decimal
is hiding

125% = 1.25

Move the decimal two places to the left. The percent was larger than 100%, so the decimal will have a whole number because it is greater than one whole.

125% = 1.25

7% = 0.07

Move the decimal two places to the left. Place a zero in the open place.

7% = 0.07

Practice Problems

Convert from a percent to a decimal:

7.26 43%

7.27 8%

7.28 64%

7.29 145%

7.30 333%

The Steps to Convert from a Percent to a Fraction

1. Write the percent as a fraction by writing the percent over 100.
2. If necessary, reduce the fraction to the lowest terms.

Examples

$45\% = \frac{45}{100}$ or $\frac{9}{20}$

Place the percent over 100 to make the fraction $\frac{45}{100}$. Then reduce the fraction by dividing both the numerator and the denominator by 5. The fraction in its simplest form is $\frac{9}{20}$.

$6\% = \frac{6}{100}$ or $\frac{3}{50}$

Place the percent over 100 to make the fraction $\frac{6}{100}$. Then reduce the fraction by dividing both the numerator and the denominator by 2. The fraction in its simplest form is $\frac{3}{50}$.

$125\% = \frac{125}{100}$ or $1\frac{1}{4}$

Place the percent over 100 to make the fraction $\frac{125}{100}$. Then reduce the fraction by dividing both the numerator and the denominator by 25. The fraction in its simplest form is $\frac{5}{4}$ or $1\frac{1}{4}$.

Practice Problems

Convert from a percent to a fraction:

7.31 29%

7.32 4%

7.33 60%

7.34 135%

7.35 150%

The Steps to Convert from a Fraction to a Decimal

There are two ways that I teach my students to convert from a fraction to a decimal. The first way is to make the fraction into an equivalent fraction with 10 or 100 in the denominator. If the fraction is not easy to do that way, don't worry. Another way is to just use a calculator!

1. Make an equivalent fraction that has 10 or 100 as the denominator. This makes it easy to change it to a decimal.
2. If the fraction cannot be changed to an equivalent fraction, then use a calculator and divide the numerator by the denominator.

Examples

$\frac{3}{5} = 0.6$

The first step is to look at the fraction and see if you can make an equivalent fraction that has a 10 or 100 in the denominator. You can change $\frac{3}{5}$ to the equivalent fraction $\frac{6}{10}$ by multiplying the numerator and denominator by 2. $\frac{6}{10}$ can be written as the decimal 0.6.

$\frac{2}{25} = 0.08$

The first step is to look at the fraction and see if you can make an equivalent fraction that has a 10 or 100 in the denominator. You can change $\frac{2}{25}$ to the equivalent fraction $\frac{8}{100}$ by multiplying the numerator and denominator by 4. $\frac{8}{100}$ can be written as the decimal 0.08.

$\frac{4}{9} = 0.44$

Looking at this fraction, you cannot make an equivalent fraction, because there is no easy number that you can multiply by 9 (the denominator) to get 10 or 100. In this case, the easiest way to convert the fraction to a decimal is to use a calculator to divide. 4 divided by 9 is 0.444444, a repeating decimal. So, using rounding, you can determine that the decimal equivalent to $\frac{4}{9}$ is 0.44.

$\frac{4}{9} = 0.4\overline{4}$ ← this shows that the decimal repeats

Practice Problems

Convert each fraction to a decimal:

7.36 $\frac{2}{5}$

7.37 $\frac{7}{20}$

7.38 $\frac{2}{50}$

7.39 $\frac{8}{9}$

7.40 $\frac{6}{8}$

The Steps to Convert from a Fraction to a Percent

1. Make an equivalent fraction that has 100 as the denominator. This makes it easy to change it to a percent.
2. If the fraction cannot be changed to an equivalent fraction, use a calculator and divide the numerator by the denominator to get a decimal. Then move the decimal two places to the right to get the percent.

Examples

$\frac{1}{5} = 20\%$

The first step is to look at the fraction and see if you can make an equivalent fraction that has a 100 as the denominator. You can change $\frac{1}{5}$ to the equivalent fraction $\frac{20}{100}$ by multiplying the numerator and denominator by 20. $\frac{20}{100}$ is 20%.

$\frac{1}{6} = 0.08$

Looking at this fraction, you cannot make an equivalent fraction because there is not an easy number that you can multiply by 6 (the denominator) to get 100. In this case, the easiest way is to use a calculator to divide. 1 divided by 6 is 0.16666666. This is a repeating decimal. So, using rounding, you can determine that the decimal equivalent to $\frac{1}{6}$ is 0.17. Move the decimal two places to the right to get 17%.

$\frac{1}{6} = 0.166\overline{6} = 17\%$ when rounded

Practice Problems

Convert each fraction to a percent:

7.41 $\frac{2}{5}$

7.42 $\frac{3}{20}$

7.43 $\frac{6}{7}$

7.44 $\frac{3}{9}$

7.45 $\frac{6}{8}$

7.46 $\frac{3}{25}$

Chapter 8

Measurement

In this chapter, you will get tips on all things measurement—perimeter, area of polygons, and volume. I will cover the formulas needed to solve problems and a few shortcuts that I use with students in my classroom. At the end of this chapter, I have included a list of the formulas for your reference.

Perimeter

Perimeter is the distance around a shape. The distance around the shape is measured in units of length—inches (in), feet (ft), yards (yd), meters (m), etc. In my classroom, I relate perimeter to being like the "fence" around the object, since it is the distance around a shape. To find the perimeter, add the lengths of all the sides. Then name the unit of length used to measure the sides.

Examples

Rectangle

The lines on the rectangle show that the top and bottom are equal in length and the right and left sides are equivalent in length. Add the lengths of all the sides together. Here, the perimeter is 20 in.

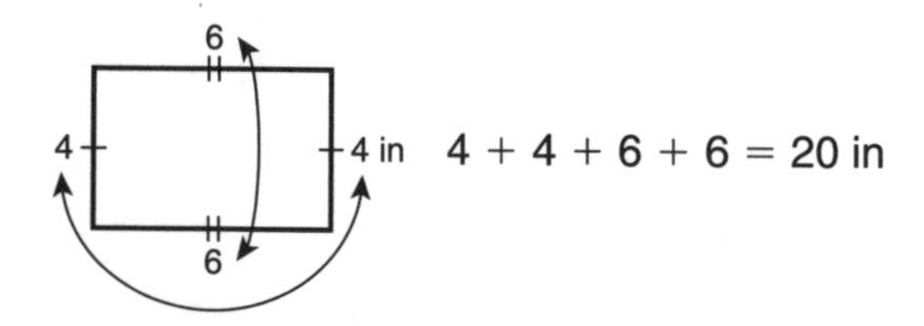

Triangle

There are three types of triangles:

- **Equilateral.** The three sides of an equilateral triangle are all the same length.
- **Isosceles.** An isosceles triangle has two sides that are the same length.
- **Scalene.** A scalene triangle has three different length sides.

This triangle is an equilateral triangle because all sides are the same length. The lines on the triangle show that all sides are equivalent. Add the lengths of all the sides together. Here, the perimeter is 15 cm.

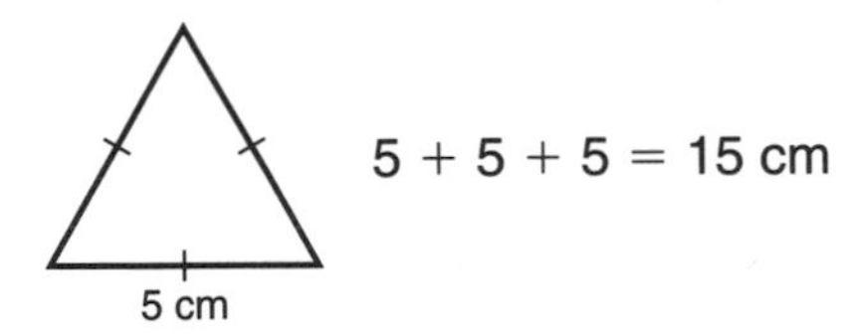

Trapezoid

Add the length of all the sides together. Here, the perimeter is 17 ft.

$3 + 3 + 4 + 7 = 17$ ft

Irregular Polygon

Add the lengths of all the sides together. Here, the perimeter is 31 yds.

$7 + 3 + 3 + 4 + 3 + 5 + 3 + 3 = 31$ yds

Practice Problems

Find the perimeter of each shape:

8.1 A playing card that has a length of 5 in and a width of 3 in

8.2 A garden that has sides with the following lengths: 10 in, 13 in, 7 in, 5 in

8.3 A pentagon (five sides) that has equal sides with a length of 9 in each

8.4 A basketball court in the shape of a rectangle that has a length of 94 ft and a width of 50 ft

8.5 A triangle that has sides of the following lengths: 52 mm, 23 mm, 15 mm

Circumference

The perimeter of a circle is different from the perimeter of a polygon because it has a special name. *Circumference* is the distance around a circle. The distance around a circle is measured in units of length—inches, feet, yards, meters, etc. The diameter of a circle is a line segment that has end points on both sides of the circle and passes through the center.

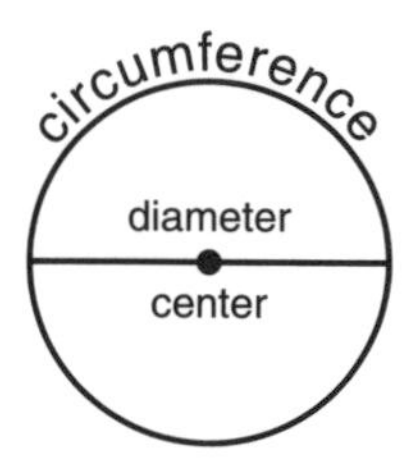

When you can identify the diameter, there is a simple formula for finding the circumference:

Circumference = $\pi(d)$, or pi × diameter

Pi (π) is roughly equal to 3.14. You can also use a calculator with a π symbol.

Examples

The diameter is pictured in the circle below. Use the formula $\pi(d)$ to find the circumference. The diameter of the circle is 7 cm, so 3.14 times 7 is 21.98. The circumference of the circle is 21.98 cm.

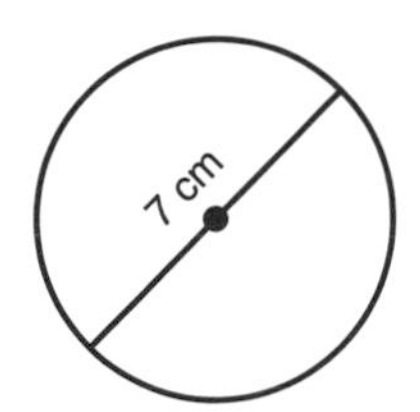

formula = π(d)
diameter = 7 cm
circumference = 3.14 (7)
circumference = 21.98 cm

The radius is pictured in the circle below. The diameter is twice the size of the radius. To figure out the diameter, multiply the radius by 2 ($5 \times 2 = 10$). The diameter of this circle is 10 in. Then, use the formula π(d) to find the circumference. The diameter of the circle is 10 in, so 3.14 times 10 is 31.4. The circumference of the circle is 31.4 in.

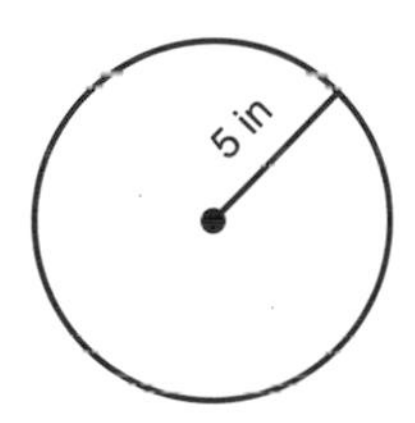

diameter= 2(r) = 2(5) = 10 in
diameter = 10 in
circumference = 3.14 (10)
circumference = 31.4 in

Practice Problems

Find the circumference of each circle:

8.6 A circle with a diameter of 8 in

8.7 A pie that has a 13-in diameter

8.8 A circle with a radius of 3.5 cm

8.9 A bicycle wheel with a diameter of 16 in

8.10 A rug with a radius of 3 ft

Area

The *area* is the amount of space taken up inside a closed shape. I relate the area being like the rug because area is the space inside a shape. Area is measured in square units. The area of a shape can be measured in any unit—cm^2, in^2, yd^2, etc. There are different formulas depending on the shape. I will give examples of common shapes, but you can also refer to the end of this chapter to see all the formulas quickly.

Examples

Rectangle

Here is the formula for finding the area of a rectangle:

$A = b \times h$

A is the area, b is the length of the base, and h is the height of the rectangle.

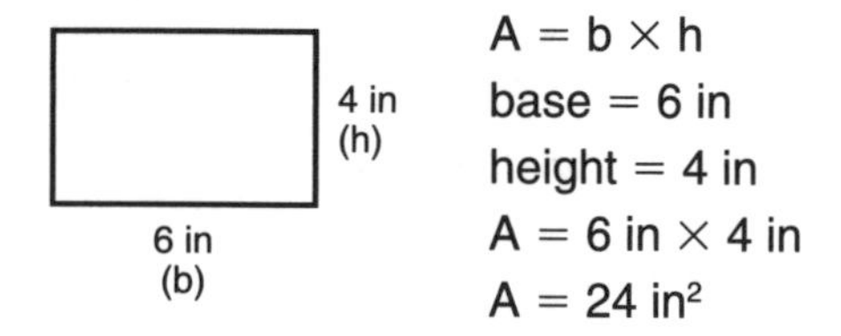

Square

Here is the formula for finding the area of a square:

$A = s^2$ or (side × side)

A is the area and s is the length of a side of the square.

5 ft

$A = s^2$ or (side × side)
side = 5 ft
A = 5 ft × 5 ft
A = 25 ft^2

Parallelogram

Here is the formula for finding the area of a parallelogram:

$A = b \times h$

A is the area, b is the length of the base, and h is the height of the parallelogram.

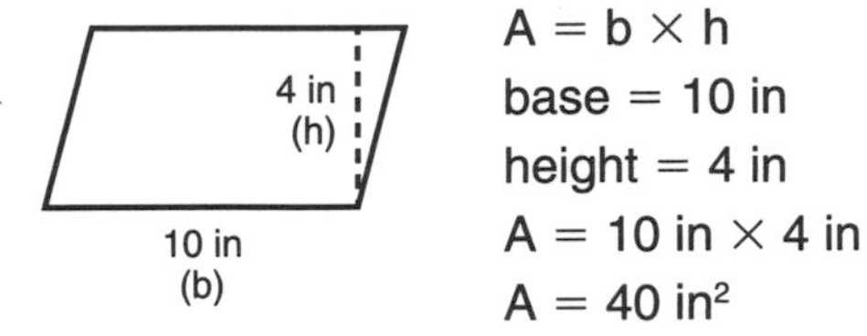

Triangle

Here is the formula for finding the area of a triangle:

$A = (b \times h) / 2 \text{ or } (b \times h) \times \frac{1}{2}$

A is the area, b is the length of the base, and h is the height of the triangle. The base and the height can look different depending on the type of triangle.

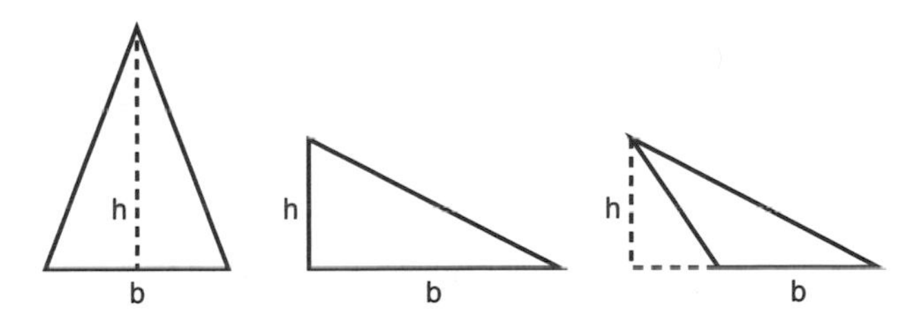

For children, the easiest way to figure out the area of a triangle is to multiply the base by the height and then divide the answer by 2.

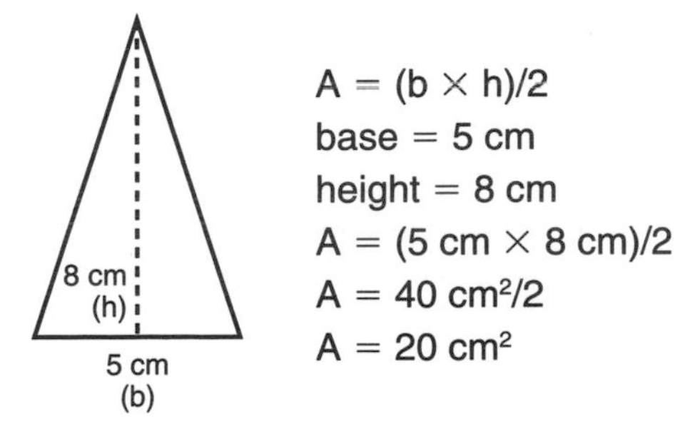

In this triangle, the height is shown by extending the base to meet the vertex at the top of the triangle. Notice that a decimal appears in the length of one of the sides. If necessary, use the algorithm for multiplication of decimals in Chapter 3, "Multiplication Algorithms."

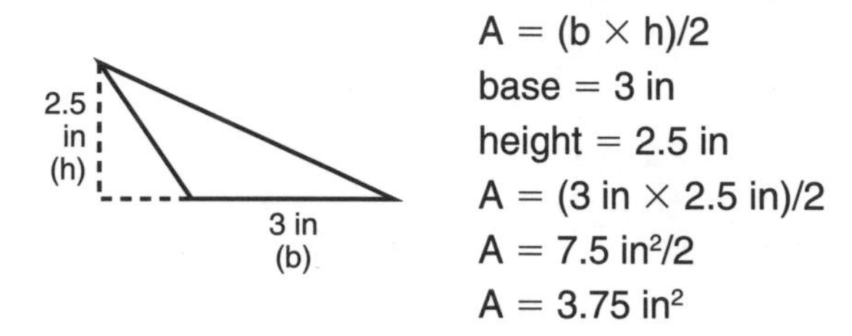

$A = (b \times h)/2$
base $= 3$ in
height $= 2.5$ in
$A = (3 \text{ in} \times 2.5 \text{ in})/2$
$A = 7.5 \text{ in}^2/2$
$A = 3.75 \text{ in}^2$

This is a right triangle, and the height is actually the length of one of the sides.

4
cm
(h)
7 cm
(b)

$A = (b \times h)/2$
base $= 7$ cm
height $= 4$ cm
$A = (7 \text{ cm} \times 4 \text{ cm})/2$
$A = 28 \text{ cm}^2/2$
$A = 14 \text{ cm}^2$

Practice Problems

Find the area of each shape:

8.11 A rectangle that has a base of 15 in and height of 3 in

8.12 A rectangle that has a base of 12.5 cm and height of 4 cm

8.13 A chess board that measures 10 in on each side

8.14 A parallelogram that has a base of 9 in and height of 10 in

8.15 A square garden with a side that measures 7 ft

8.16
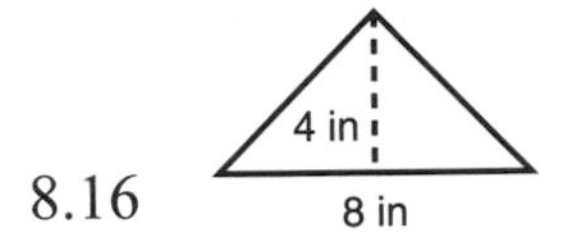

8.17
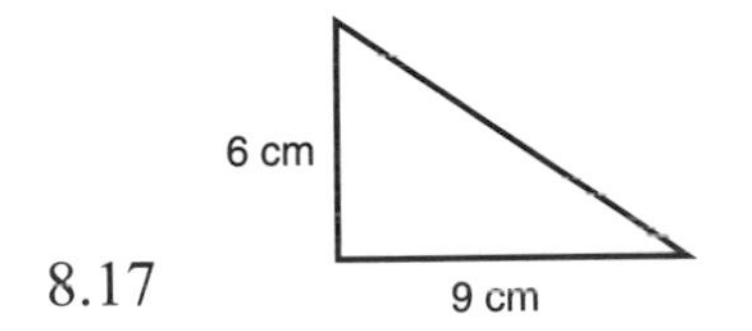

8.18
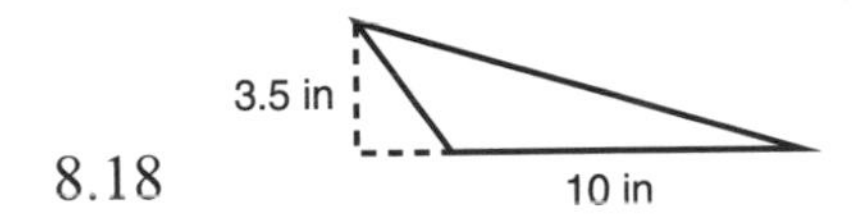

8.19 The area of a triangular garden with a base of 13 ft and height of 5 ft

8.20 The area of a triangular rug with a base of 8 ft and height of 5.5 ft

Area of a Circle

The formula for the area of a circle uses the radius of the circle. The *radius* is a line segment that connects the center to any point on the circle. What happens if you aren't given the radius of the circle? If given the diameter, you can find the radius of a circle with the following formula:

Area of a circle = $\pi(r^2)$ or pi × (radius squared)

Pi (π) is equal to 3.14. You can also use a calculator with a π symbol.

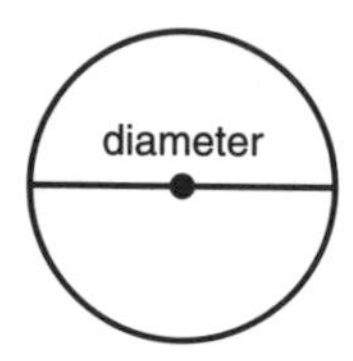

radius = 1/2 (diameter)
or diameter/2

Examples

The radius is pictured in the following circle. Use the formula $\pi(r^2)$ to find the area of the circle. The radius of the circle is 3 cm, so $3^2 = 9$. The formula for area is $\pi(r^2)$, so 3.14(9) = 28.26. The area of the circle is 28.26 cm^2.

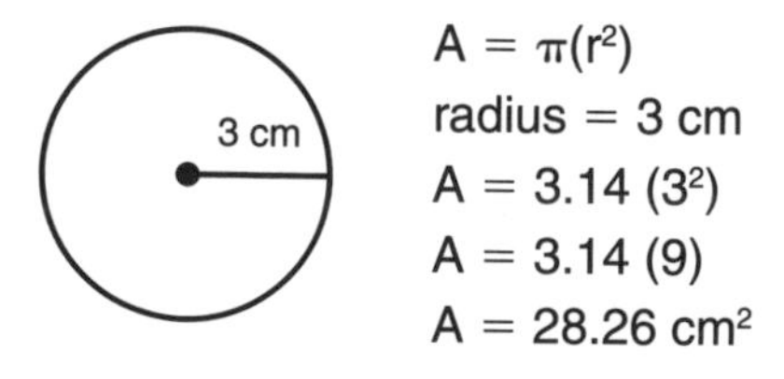

The diameter is pictured in the following circle. To find the radius, divide the diameter by 2. If the diameter is 14 cm, then the radius is 7 cm. The radius of the circle is 7 cm, so $7^2 = 49$. The formula for area is $\pi(r^2)$, so 3.14(49) = 153.86. The area of the circle is 153.86 cm^2.

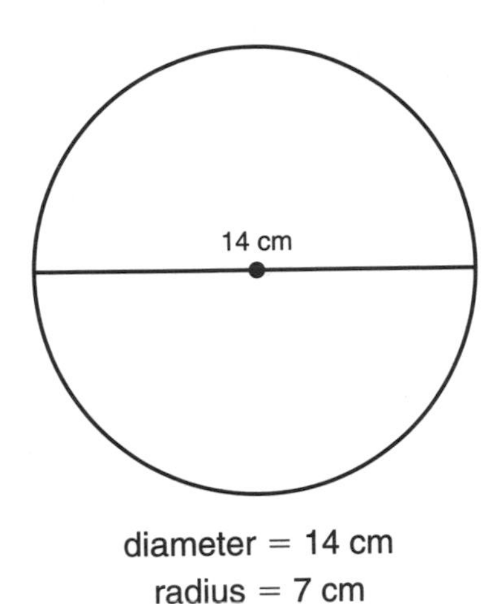

diameter = 14 cm
radius = 7 cm

$A = \pi(r^2)$
radius = 7 cm
$A = 3.14\ (7^2)$
$A = 3.14\ (49)$
$A = 153.86\ cm^2$

Practice Problems

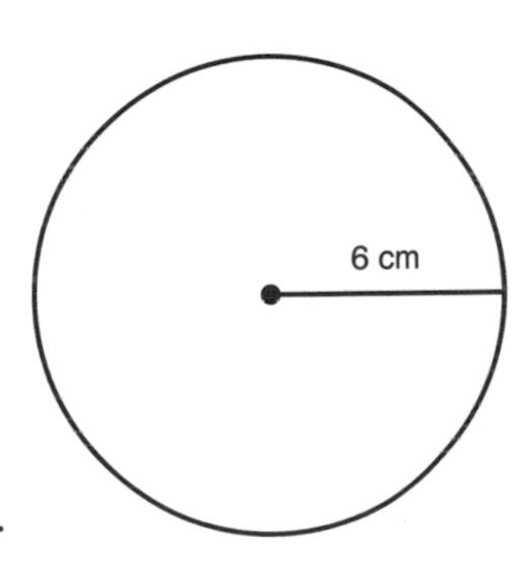

8.21 Find the radius of the circle.

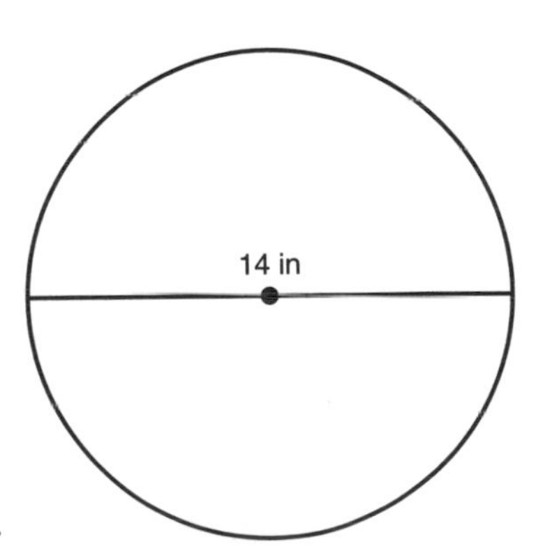

8.22 Find the radius of the circle.

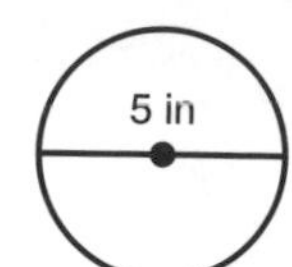

8.23 Find the radius of the circle.

8.24 Find the area of a circular table with a radius of 2 ft.

8.25 Find the area of a circular rug with a diameter of 15 ft.

Volume of a Solid Figure

Just like area, there are different formulas used to find volume depending on the shape of the geometric solid. To find the volume of geometric solids, you find the area of the base and then multiply that by the height. A solid foundation of formulas for area is necessary to find the volume. Use the aforementioned formulas for area (which also appear at the end of the chapter) to help you with volume.

Examples

Rectangular Prism

The formula for the volume of a rectangular prism (where the base is a rectangle) is as follows:

$V = B \times h$

V is the volume, B is the area of the rectangular base, and h is the height of the prism.

For the example shown next, the first step is to find the area of the base. The base is a rectangle, so to find the area of the base, multiply 4 cm by 2 cm. The area of the base of the prism is 8 cm^2. Next, multiply this area by the height of the prism: 8 $cm^2 \times 6$ cm. The volume of the prism is 48 cm^3.

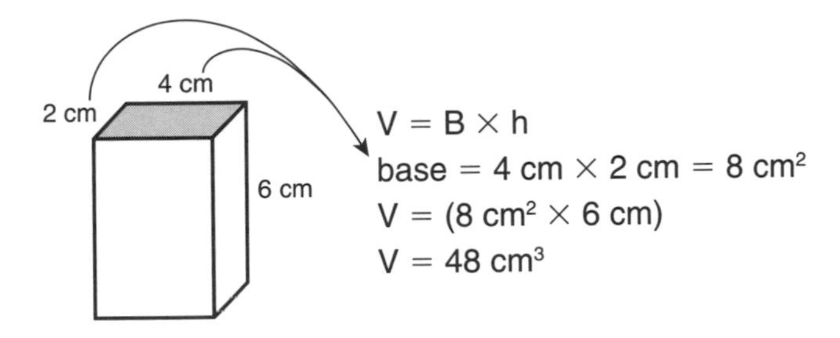

For the example shown next, the first step is to find the area of the base. The base is a rectangle, so to find the area of the base, multiply 10 cm by 2 cm. The area of the base of the prism is 20 cm^2. Next, multiply this area by the height of the prism: 20 $cm^2 \times 4$ cm. The volume of the prism is 80 cm^3.

$V = B \times h$

base = 10 cm × 2 cm = 20 cm^2

V = (20 cm^2 × 4 cm)

V = 80 cm^3

Triangular Prism

The formula for the volume of a triangular prism (where the base is a triangle) is as follows:

$V = B \times h$

V is the volume, B is the area of the triangular base, and h is the height of the prism.

For the example shown next, the first step is to find the area of the base. The base is a triangle, so to find the area of the base, multiply 4 in by 3 in and divide the product by 2. The area of the base of the prism is 6 in^2. Next, multiply this area by the height of the prism: 6 in^2 × 7 in. The volume of the prism is 42 in^3.

3 in
7 in
4 in

$V = B \times h$

base = 6 in^2

$V = (6 \text{ in}^2 \times 7 \text{ in})$

$V = 42 \text{ in}^3$

area of a triangle:

$$\frac{b \times h}{2} = \frac{4 \times 3}{2} = \frac{12}{2} = 6 \text{ in}^2$$

For the example shown next, the first step is to find the area of the base. The base is a triangle, so to find the area of the base, multiply 5 cm by 3 cm and divide the product by 2. The area of the base of the prism is 7.5 cm^2. Next, multiply this area by the height of the prism: 7.5 cm^2 × 10 cm. The volume of the prism is 75 cm^3.

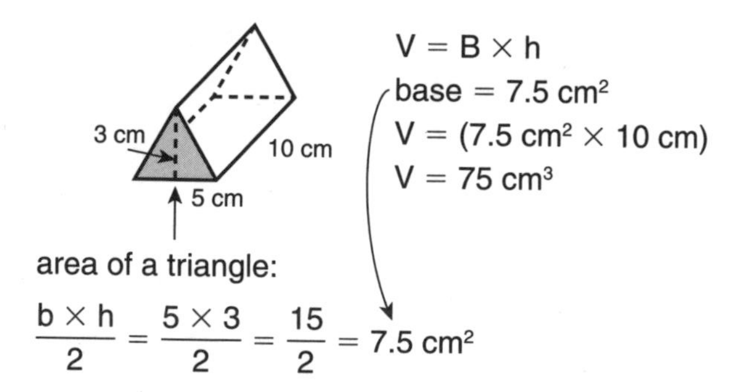

Practice Problems

8.26 Find the volume of the rectangular prism.

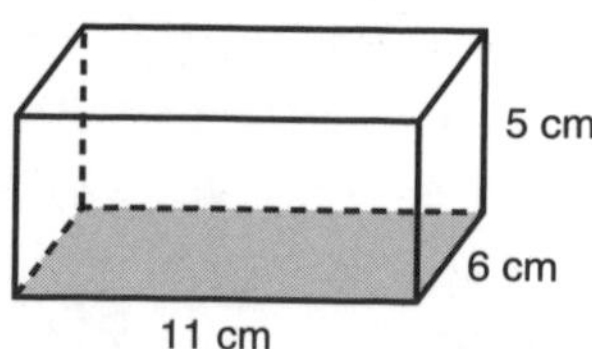

8.27 Find the volume of the rectangular prism.

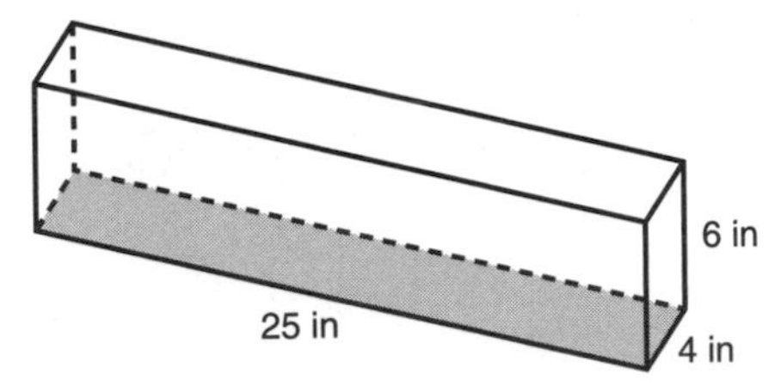

8.28 Find the volume of the triangular prism.

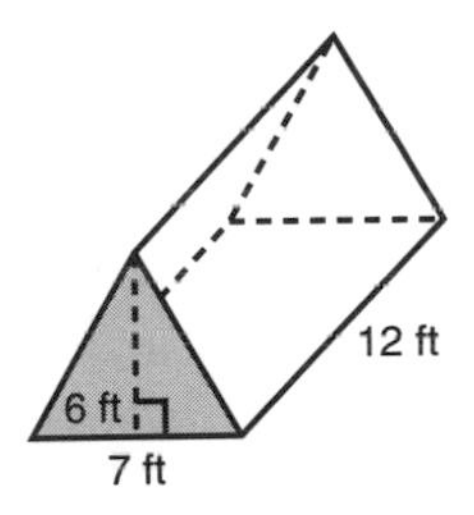

8.29 Find the volume of a box with a base area of 24 in^2 and a height of 8 inches.

8.30 Find the volume of a triangular prism with a base area of 20 in^2 and a height of 12 inches.

Volume of a Cylinder and Cone

A cylinder and a cone both have a circle as the base of the solid figure. For this reason, they both require finding the area of the circle prior to completing the rest of the formula. To find the area of a circle, use the formula $A = \pi(r^2)$, or pi × (radius squared). If you are unsure of this formula, refer to the section "Area of a Circle."

Examples

Cylinder

The formula for the volume of a cylinder (where the base is a circle) is as follows:

$V = B \times h$

V is the volume, B is the area of the circle base, and h is the height of the cylinder.

For the example shown next, the first step is to find the area of the base. The base is a circle, so to find the area of the base, multiply pi (3.14) by 4^2 (16). The area of the base of the cylinder is 50.24 cm^2. Next, multiply the base area by the height of the prism: 8 cm × 50.24 cm^2. The volume of the cylinder is 401.92 cm^3.

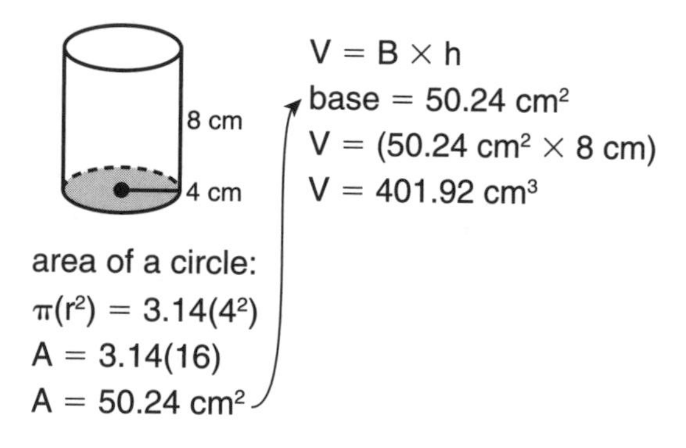

For the example shown next, the first step is to find the area of the base. The diameter is given, so divide the diameter by 2 to get the radius. The radius is 5 in. The base is a circle, so to find the area of the base, multiply pi (3.14) by 5^2 (25). The area of the base of the cylinder is 78.5 in^2. Next, multiply this area by the height of the prism: 4 in × 78.5 in^2. The volume of the cylinder is 314 in^3.

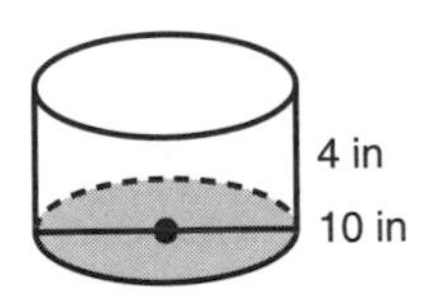

$V = B \times h$
base $= 78.5\ in^2$
$V = (78.5\ in^2 \times 4\ in)$
$V = 314\ in^3$

area of a circle:
$\pi(r^2) = 3.14(5^2)$
$A = 3.14(25)$
$A = 78.5\ in^2$

Cone

The formula for the volume of a cone (where the base is a circle) is as follows:

$V = (B \times h)\ \frac{1}{3}$ or (area of the base × height) / 3

V is the volume, B is the area of the circle base, and h is the height of the cone.

For the example shown next, the first step is to find the area of the base. The base is a circle, so to find the area of the base, multiply pi (3.14) by 3^2 (9). The area of the base of the cone is 28.26 cm^2. Next, multiply this area by the height of the prism: 8 cm × 28.26 cm^2. Finally, multiply the answer by $\frac{1}{3}$, or divide by 3. The volume of the cone is 75.36 cm^3.

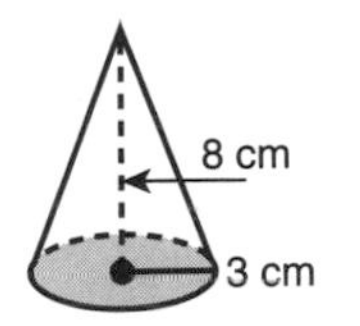

$V = (B \times h)\frac{1}{3}$
base $= 28.26\ cm^2$
$V = (28.26\ cm^2 \times 8\ cm) \times \frac{1}{3}$
$V = 75.36\ cm^3$

area of a circle:
$\pi(r^2) = 3.14(3^2)$
$A = 3.14(9)$
$A = 28.26\ cm^2$

Practice Problems

8.31 Find the volume of the cylinder.

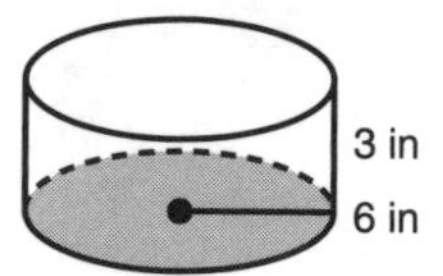

8.32 Find the volume of the cylinder.

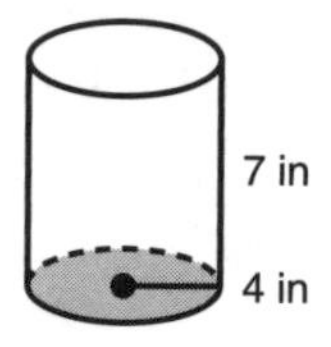

8.33 Find the volume of the cone.

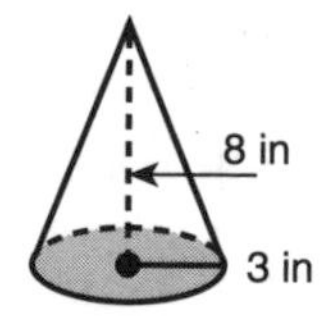

8.34 Find the volume of an ice cream cone with a radius of 3 cm and a height of 10 cm.

8.35 Find the volume of a birthday hat with a diameter of 7 cm and a height of 15 cm.

Quick Formula Reference for Perimeter, Area, and Volume

Perimeter

Labeled in units—in, cm, ft, yd, etc.

- **Perimeter:** Add up the lengths of all the sides
- **Circumference:** $\pi(d)$ or pi × diameter

$\pi(\text{pi}) = 3.14$

Area

Labeled in square units—in^2, cm^2, ft^2, yd^2, etc.

- **Area of a rectangle:** $b \times h$
- **Area of a square:** s^2 or side × side
- **Area of a parallelogram:** $b \times h$
- **Area of a triangle:** $(b \times h) / 2$ or $(b \times h) \times \frac{1}{2}$
- **Area of a circle:** $\pi(r^2)$ or pi × (radius squared)

Volume

Labeled in cubic units—in^3, cm^3, ft^3, yd^3, etc.

- **Volume of a rectangular prism (the base is a rectangle):** $B \times h$
- **Volume of a triangular prism (the base is a triangle):** $B \times h$
- **Volume of a cylinder (the base is a circle):** $B \times h$
- **Volume of a cone (the base is a circle):** $(B \times h)\ \frac{1}{3}$ or (area of the base × height) / 3

Chapter 9

Algebra Concepts

Algebra is a mathematical concept that uses numbers as well as letters. The letter or letters in an algebra equation are used to represent an unknown number. This letter in the equation is called a *variable*.

To solve an equation, you use the numbers in the "number sentence" to find the correct solution. For example in the number sentence $6 + x = 15$, the variable x stands for an unknown number. In this equation, x equals 9.

In this chapter, the multiplication symbol (×) will be replaced by an asterisk (*) because in algebra, an *x* is used to represent an unknown number.

Parentheses

In algebra, there are instances in an equation when there are numerous operations used—addition, subtraction, multiplication, and/or division. When solving, it might be confusing to decide which operation to compute first to get to the correct solution. For example, in the equation $5 + 4 * 8 = x$, you can obtain two different solutions depending on the order in which you complete the operations. That is, if you add 5 and 4 first and then multiply the sum by 8, the answer would be 72. But if you multiply 4 by 8 first and then add 5, the answer would be 37. Placing parentheses in the equation will make the steps to the solution clear. The items in parentheses are always computed first in an equation. In the examples that follow, you will see that the placement of the parentheses drives how the problem is computed.

Examples

In some problems, the equation might be solved for x. In others, you need to add parentheses to the equation to make the answer true.

Solve for x:

$5 + 4 * 8 = x$

$(5 + 4) * 8 = x$

In this equation, add 5 and 4 first, and then multiply by 8. The solution is $x = 72$.

Solve for x:

$5 + (4 * 8) = x$

In this equation, multiply 4 and 8 first and then add 5. The solution is $x = 37$.

$(50 - 6) \div 2 = x$

In this equation, subtract 50 and 6, and then divide by 2. The solution is $x = 22$.

$x = (16 - 8) + (5 * 6)$

In this equation, there are two sets of parentheses. Solve both problems in the parentheses and then add the solutions together. Add 8 and 30 to get the solution, 38.

Add parentheses to make the equation correct:

$74 = 100 - 30 + 4$

In this problem, add the parentheses to make the equation correct. You need to place the parentheses around $100 - 30$ because this equals 70, and $70 + 4 = 74$. This makes the equation a true number sentence: $74 = (100 - 30) + 4$.

$35 + 25 \div 5 = 40$

In this number sentence, you must place the parentheses around the $25 \div 5$ because this equals 5, and $5 + 35 = 40$. This makes the equation a true number sentence: $35 + (25 \div 5) = 40$.

Practice Problems

Solve for x.

9.1 $(9 * 9) - 10 = x$

9.2 $7 + (36 \div 6) = x$

9.3 $(5 * 4) - (3 * 4) = x$

9.4 $x = 100 - (5 * 7)$

9.5 $x = (100 \div 4) + 25$

Place parentheses to make the equation true.

9.6 $5 + 7 * 12 = 144$

9.7 $3 * 6 - 9 = 9$

9.8 $72 \div 9 * 8 = 64$

9.9 $18 - 3 * 4 = 60$

9.10 $22 = 6 * 2 + 5 * 2$

Order of Operations

Sometimes, an equation may have parentheses that tell you which part of the equation to solve first. Other times, however, there may not be parentheses. In these cases, there is an order of operations for solving the equation. You can remember the order of operations using the following mnemonic saying: Please Excuse My Dear Aunt Sally (PEMDAS). It stands for the following:

- **P:** Parentheses
- **E:** Exponents
- **M:** Multiplication
- **D:** Division
- **A:** Addition
- **S:** Subtraction

So, first do any operations that are inside the parentheses. Next, calculate all the exponents in the expression. Then, multiply and divide in order of the expression from left to right. Finally, add and subtract in order from left to right.

Examples

$(4 * 3) + 5^2$

Following PEMDAS, begin by multiplying 4 by 3, because it is inside the parentheses. Then compute the exponent, 5^2 equals 25. Finally, add 12 and 25 for a final answer of 37.

$\mathbf{(4 * 3)} + 5^2$ (**P**arentheses)

$12 + \mathbf{5^2}$ (**E**xponents)

$\mathbf{12 + 25} = 37$ (**A**dd/**S**ubtract left to right)

$6^2 – 8 \div 2 * 6$

There are no parentheses in this problem, so compute the exponent first: $6^2 = 36$. Then multiply/divide in order from left to right: 8 ÷ 2 = 4, and 4 * 6 = 24. Finally, subtract 24 from 36 for a final answer of 12.

$\mathbf{6^2}$ – 8 ÷ 2 * 6	(**E**xponents)
36 – **8 ÷ 2** * 6	(**M**ultiply/**D**ivide left to right)
36 – **4 * 6**	(**M**ultiply/**D**ivide left to right)
36 – 24 = 12	(**A**dd/**S**ubtract left to right)

5 * (3 – 1) + 4 – 5

Begin by subtracting 1 from 3 because it is inside parentheses. Then multiply/divide in order from left to right: 5 * 2 = 10. Finally, add/subtract in order from left to right: 10 + 4 = 14; 14 – 5 equals 9. This is the final answer.

5 * **(3 – 1)** + 4 – 5	(**P**arentheses)
5 * 2 + 4 – 5	(**M**ultiply/**D**ivide left to right)
10 + 4 – 5 = 9	(**A**dd/ **S**ubtract left to right)

(28 – 6 * 3) + 5 – 7

Notice in this problem that there are multiple operations within the parentheses. In this case, follow the same rules for PEMDAS when computing inside the parentheses. Begin in the parentheses by multiplying 6 by 3: 6 * 3 = 18. Then subtract 18 from 28: 28 – 18 = 10. Getting rid of the parentheses leaves 10 + 5 – 7. The last step is to add/subtract in order from left to right. The final answer is 8.

(28 – **6 * 3**) + 5 – 7	(**P**arentheses) (**M**ultiply/**D**ivide left to right)
(28 – 18) + 5 – 7	(**P**arentheses) (**A**dd/**S**ubtract left to right)
10 + 5 – 7 = 8	(**A**dd/**S**ubtract left to right)

Practice Problems

Use PEMDAS to evaluate each expression.

9.11 $6 * 5 + 4$

9.12 $(6 + 2) * 4^2$

9.13 $(3 * 4 + 8) \div 5$

9.14 $7^2 + 8 \div 8 - 15$

9.15 $(100 \div 2^2) - (3^2 + 4)$

Algebraic Expressions

Sometimes in algebraic expressions, you will be given a value for the variable. For example, a problem might ask you to evaluate the expression with the variables given. In this instance, you might know that $x = 4$ and $y = 3$. In other problems, you might be asked to write the phrase in algebraic terms using a variable for an unknown number. A variable can be represented by any letter, and the value of the variable could be a whole number, positive number, negative number, or fraction.

Examples

Table 9.1 shows several expressions rewritten in algebraic terms. (Remember: the variable, or unknown number, can be represented by any letter.)

Table 9.1 Examples of Ordinary Language Changed to an Algebraic Expression

Ordinary Language	Algebraic Expression
a certain number plus 4	$x + 4$
eight less than a certain number	$x - 8$
two less than twice a certain number	$2x - 2$
Alex is three years older than Sam.	$s + 3$ (s represents Sam's age)
The cost of a pack of gum less 30 cents.	$g - 30$ (g represents the cost of the gum)

Following are examples of evaluating an expression using substitution. Here, you will be given the value for a variable and substitute that variable in the expression. For each expression, let $x = 5$ and $y = 4$.

$(x + y) \div 3$

$(5 + 4) \div 3$

$9 \div 3 = 3$

$5x + 3y$

$5(5) + 3(4)$

$25 + 12 = 37$

$x^2 - y^2$

$5^2 - 4^2$

$25 - 16 = 9$

Practice Problems

Rewrite as an algebraic expression.

9.16 a certain number divided by 2

9.17 2 times a certain number minus 4

9.18 a certain number to the second power plus 5

9.19 25 more than the number of pairs of shoes owned

9.20 $5 per hour worked in a week plus $10

Substitute each expression. Let $x = 6$, $y = 10$ and $z = 15$.

9.21 $(5x) + (3y)$

9.22 $y^2 + 15$

9.23 $(x * y) - z$

9.24 $(45 \div z) + (x * y)$

9.25 $y + 2(x + z)$

Appendix A

Answers to Practice Problems

Chapter 1: Addition Algorithms

1.1 668
1.2 722
1.3 8,927
1.4 3,330
1.5 17,612
1.6 668
1.7 698
1.8 799
1.9 588
1.10 899
1.11 1,337
1.12 2,514
1.13 14,085
1.14 38,110
1.15 130,268
1.16 1,260
1.17 887
1.18 2,106
1.19 43,545
1.20 72,700
1.21 11
1.22 4
1.23 83.9
1.24 81.61
1.25 0.576
1.26 9.96
1.27 1.22
1.28 11.567
1.29 10.32
1.30 28.12

Chapter 2: Subtraction Algorithms

2.1 503
2.2 417
2.3 190
2.4 4,388
2.5 2,064
2.6 49
2.7 513
2.8 438
2.9 2,546
2.10 2,873
2.11 64
2.12 89
2.13 578
2.14 4,034

2.15 5,606
2.16 211
2.17 51
2.18 711
2.19 4,049
2.20 4,081
2.21 4.4
2.22 8.3
2.23 22.29
2.24 2.317
2.25 6.12
2.26 13.8
2.27 9.58
2.28 60.476
2.29 272.44
2.30 0.364

Chapter 3: Multiplication Algorithms

3.1 1,904
3.2 44,712
3.3 21,672
3.4 12,062
3.5 2,369,026
3.6 1,560
3.7 5,984
3.8 13,815
3.9 43,825
3.10 1,746,026
3.11 9.92
3.12 855.6
3.13 448.8
3.14 18.72
3.15 29.184
3.16 5,428
3.17 9,180
3.18 330,363
3.19 1,446,505
3.20 266,015
3.21 1,702

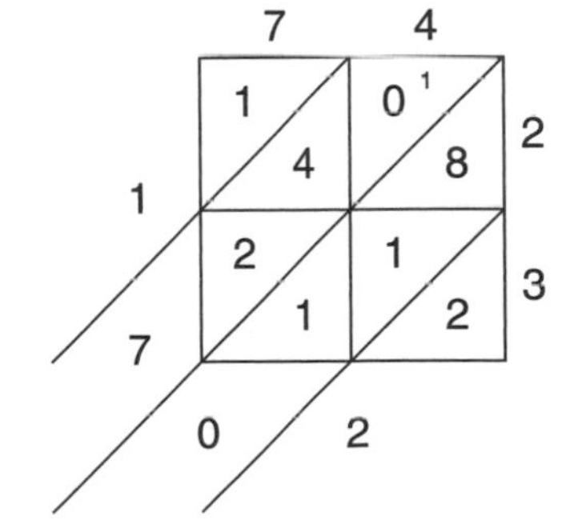

3.22 37,228

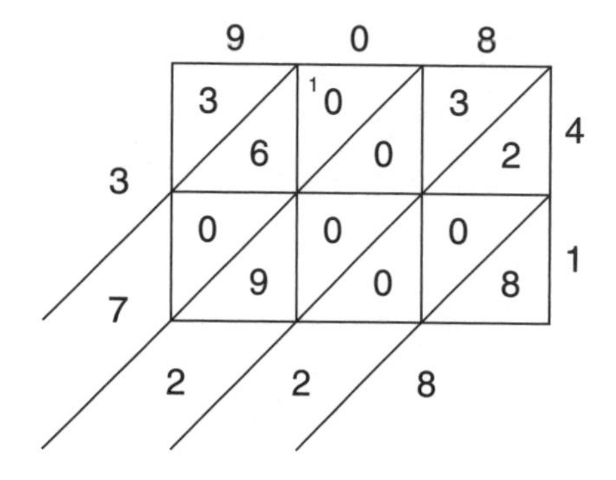

3.23 605,728

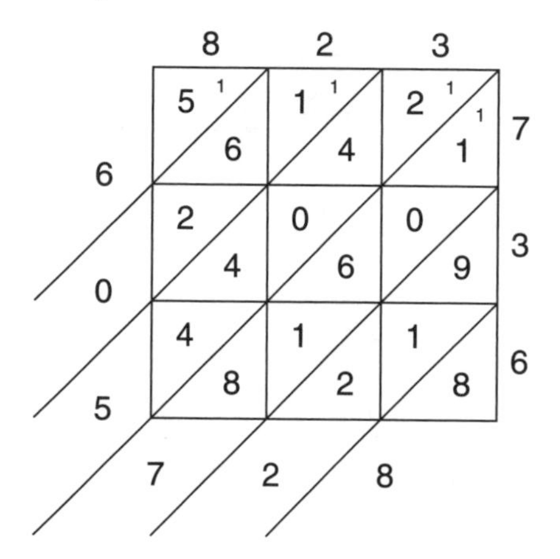

3.24 1,566,860

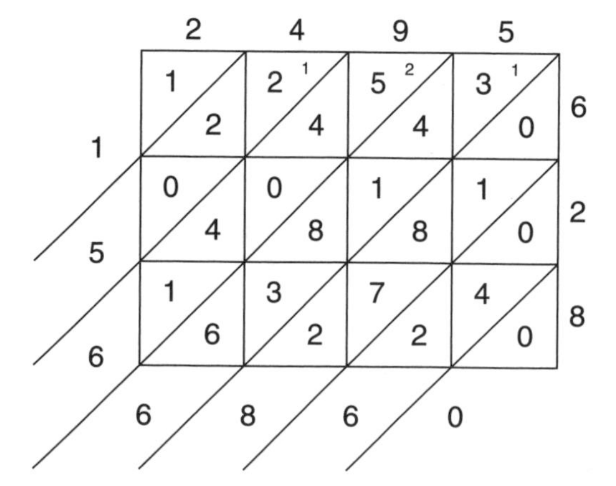

3.25 291,252

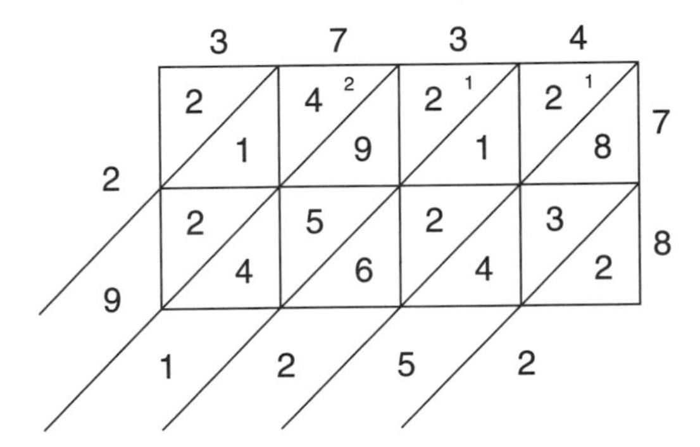

3.26 27.69

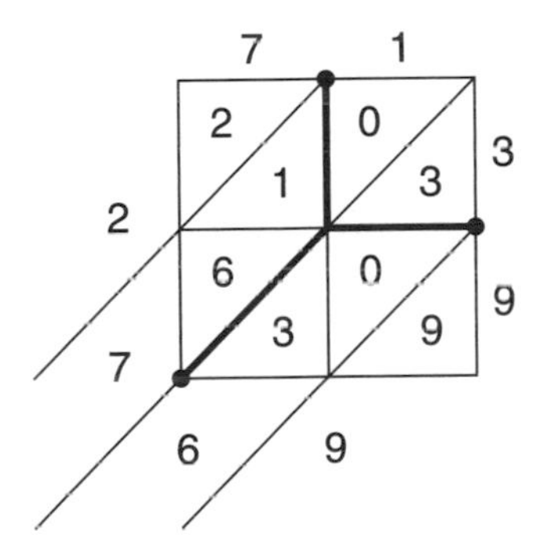

3.27 14.348

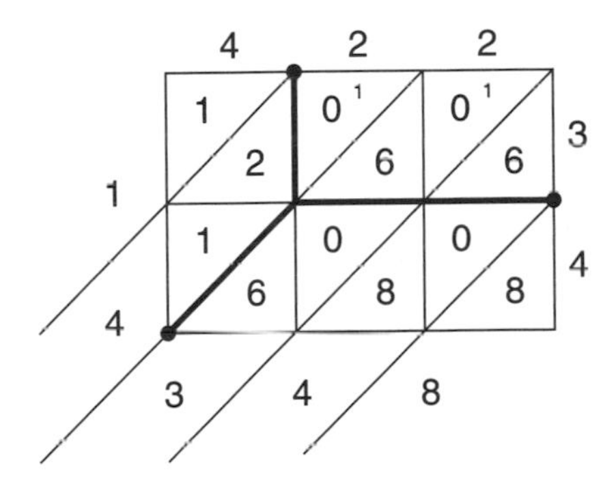

3.28 596.3

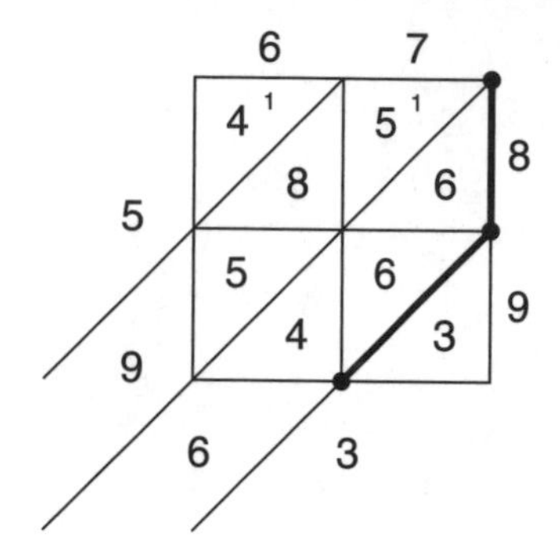

3.29 1,140

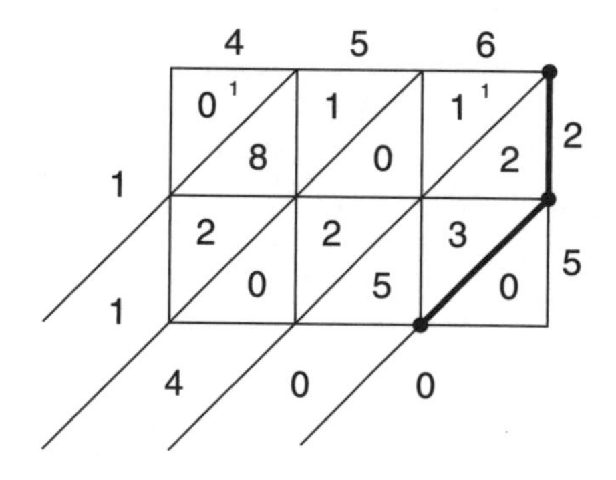

3.30 310.64

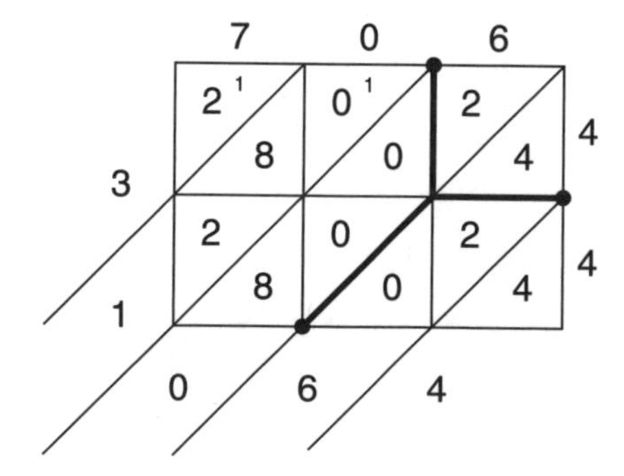

Chapter 4: Division Algorithms

4.1 142

4.2 181 r4 or $181\frac{1}{2}$

4.3 480 r3 or $480\frac{1}{4}$

4.4 40 r16 or $40\frac{2}{3}$

4.5 144 r5 or $144\frac{1}{9}$

4.6 3.8

4.7 8.4

4.8 4.9

4.9 0.52

4.10 7.32

4.11 150 r2 or $150\frac{2}{5}$

4.12 486 r5 or $486\frac{5}{6}$

4.13 379 r5 or $379\frac{5}{8}$

4.14 285 r12 or $285\frac{4}{5}$

4.15 209 r33 or $209\frac{33}{35}$

Chapter 5: Fraction Concepts

5.1 $6\frac{1}{2}$

5.2 $3\frac{3}{5}$

5.3 $2\frac{7}{8}$

5.4 $7\frac{6}{7}$

5.5 $6\frac{8}{9}$

5.6 $\frac{23}{4}$

5.7 $\frac{65}{9}$

5.8 $\frac{38}{3}$

5.9 $\frac{78}{9}$

5.10 $\frac{103}{11}$

5.11 $\frac{1}{3}, \frac{4}{12}, \frac{6}{18}$

5.12 $\frac{10}{18}, \frac{15}{27}, \frac{20}{36}$

5.13 $\frac{14}{16}, \frac{21}{24}, \frac{28}{32}$

5.14 $\frac{2}{3}, \frac{12}{18}, \frac{18}{27}$

5.15 $\frac{4}{10}, \frac{6}{15}, \frac{8}{20}$

5.16 $x = 32$

Answers may vary for 5.11–5.15

5.17 $x = 12$

5.18 $x = 12$

5.19 $x = 15$

5.20 $x = 72$

5.21 $\frac{2}{3}$

5.22 $\frac{2}{3}$

5.23 $\frac{1}{8}$

5.24 $\frac{1}{5}$

5.25 $\frac{1}{5}$

5.26 $<$

5.27 $>$

5.28 $<$

5.29 $=$

5.30 $<$

5.31 $\frac{2}{6}, \frac{4}{6}, \frac{5}{6}$

5.32 $\frac{5}{15}, \frac{5}{6}, \frac{5}{8}$

5.33 $\frac{3}{14}, \frac{2}{7}, \frac{4}{7}$

5.34 $\frac{1}{8}, \frac{7}{16}, \frac{3}{4}$

5.35 $\frac{1}{15}, \frac{2}{3}, \frac{4}{5}$

Chapter 6: Fraction Computation

6.1 $\frac{7}{8}$

6.2 $\frac{7}{10}$

6.3 $\frac{5}{7}$

6.4 $\frac{6}{8}$ or $\frac{3}{4}$

6.5 $\frac{5}{15}$ or $\frac{1}{3}$

6.6 $\frac{7}{8}$

6.7 $\frac{7}{14}$ or $\frac{1}{2}$

6.8 $\frac{12}{8}$ or $1\frac{4}{8}$ or $1\frac{1}{2}$

6.9 $\frac{12}{9}$ or $1\frac{3}{9}$ or $1\frac{1}{3}$

6.10 $\frac{58}{35}$ or $1\frac{23}{35}$

6.11 6

6.12 $9\frac{1}{9}$

6.13 $9\frac{1}{10}$

6.14 $7\frac{3}{14}$

6.15 $13\frac{9}{15}$ or $13\frac{3}{5}$

6.16 $\frac{3}{6}$ or $\frac{1}{2}$

6.17 $\frac{4}{9}$

6.18 $\frac{6}{12}$ or $\frac{1}{2}$

6.19 $\frac{5}{16}$

6.20 $\frac{4}{20}$ or $\frac{1}{5}$

6.21 $\frac{2}{12}$ or $\frac{1}{6}$

6.22 $\frac{2}{8}$ or $\frac{1}{4}$

6.23 $\frac{1}{12}$

6.24 $\frac{4}{10}$ or $\frac{2}{5}$

6.25 $\frac{2}{15}$

6.26 $3\frac{5}{12}$

6.27 $2\frac{7}{12}$

6.28 $3\frac{13}{30}$

6.29 $3\frac{8}{15}$

6.30 $4\frac{14}{20}$ or $4\frac{7}{10}$

6.31 $\frac{8}{40}$ or $\frac{1}{5}$

6.32 $\frac{12}{54}$ or $\frac{2}{9}$

6.33 $\frac{8}{80}$ or $\frac{1}{10}$

6.34 $\frac{48}{192}$ or $\frac{1}{4}$

6.35 $\frac{42}{96}$ or $\frac{7}{16}$

6.36 $8\frac{2}{4}$ or $8\frac{1}{2}$

6.37 $3\frac{2}{5}$

6.38 6

6.39 7

6.40 8

6.41 $1\frac{2}{4}$ or $1\frac{1}{2}$

6.42 $2\frac{1}{4}$

6.43 $\frac{2}{21}$

6.44 $\frac{5}{36}$

6.45 $\frac{10}{12}$ or $\frac{5}{6}$

6.46 $3\frac{8}{9}$

6.47 3

6.48 2

6.49 1

6.50 $3\frac{2}{9}$

Chapter 7: Decimals, Fractions, and Percents

7.1 $<$

7.2 $>$

7.3 $<$

7.4 $<$

7.5 $<$

7.6 4.58; 4.2; 3.41; 0.5

7.7 7.44; 6.31; 6.13; 6.014

7.8 12.99; 12.81; 12.7; 12.5

7.9 0.41; 0.52; 0.8; 1.8

7.10 3.1; 3.54; 3.74; 3.852

7.11 23.5

7.12 149

7.13 74.3

7.14 52.85

7.15 4.74

7.16 $\frac{6}{10}$ or $\frac{3}{5}$

7.17 $\frac{54}{100}$ or $\frac{27}{50}$

7.18 $\frac{80}{100}$ or $\frac{4}{5}$

7.19 $1\frac{23}{100}$

7.20 $2\frac{6}{10}$ or $2\frac{3}{5}$

7.21 40%

7.22 29%

7.23 46%

7.24 254%

7.25 160%

7.26 0.43

7.27 0.08

7.28 0.64

7.29 1.45

7.30 3.33

7.31 $\frac{29}{100}$

7.32 $\frac{4}{100}$ or $\frac{1}{25}$

7.33 $\frac{60}{100}$ or $\frac{3}{5}$

7.34 $\frac{135}{100}$ or $1\frac{35}{100}$ or $1\frac{7}{20}$

7.35 $\frac{150}{100}$ or $1\frac{50}{100}$ or $1\frac{1}{2}$

7.36 0.40

7.37 0.35

7.38 0.04

7.39 0.89

7.40 0.75

7.41 40%

7.42 15%

7.43 86%

7.44 33%

7.45 75%

7.46 12%

Chapter 8: Measurement

8.1 16 in

8.2 35 in

8.3 45 in

8.4 288 ft

8.5 90 mm

8.6 25.12 in

8.7 40.82 in

8.8 21.98 cm

8.9 50.24 in

8.10 18.84 ft

8.11 45 in^2

8.12 50 cm^2

8.13 100 in^2

8.14 90 in^2

8.15 49 ft^2

8.16 16 in^2

8.17 27 cm^2

8.18 17.5 in^2

8.19 32.5 ft^2

8.20 22 ft^2

8.21 113.04 cm^2

8.22 153.86 in^2

8.23 19.625 in^2

8.24 12.56 ft^2

8.25 176.625 ft^2

8.26 330 cm^3

8.27 600 in^3

8.28 252 ft^3

8.29 192 in^3

8.30 240 in^3

8.31 339.12 in^3

8.32 351.68 in^3

8.33 75.36 in^3

8.34 94.2 cm^3

8.35 192.325 cm^3

Chapter 9: Algebra Concepts

9.1 $x = 71$

9.2 $x = 13$

9.3 $x = 8$

9.4 $x = 65$

9.5 $x = 50$

9.6 $(5 + 7) * 12 = 144$

9.7 $(3 * 6) - 9 = 9$

9.8 $(72 \div 9) * 8 = 64$

9.9 $(18 - 3) * 4 = 60$

9.10 $22 = (6 * 2) + (5 * 2)$

9.11 34

9.12 128

9.13 4

9.14 35

9.15 12

9.16 $a \div 2$ or $[a/2]$

9.17 $2x - 4$

9.18 $x^2 + 5$

9.19 $s + 25$

9.20 \$5h + \$10

9.21 60

9.22 115

9.23 45

9.24 63

9.25 252